Contents

iii

Contents

Refugees, Environment and Development

Richard Black
University of Sussex

LONGMAN

Addison Wesley Longman Limited
Edinburgh Gate
Harlow
Essex   CM20 2JE
United Kingdom
and Associated Companies throughout the world

Published in the United States of America
by Addison Wesley Longman, New York

First published 1998

ISBN   0 582 31564 6

Visit Addison Wesley Longman on the world wide web at
http://www.awl-he.com

British Library Cataloguing-in-Publication Data
A catalogue record for this book is available from the British Library

Library of Congress Cataloging-in-Publication Data
A catalog record for this book is available from the Library of Congress

Set by 35 in 10/11pt Palatino
Produced by Addison Wesley Longman Singapore (Pte) Ltd,
Printed in Singapore

Preface

On 30 March 1995, Tanzania closed its borders to around 70 000 Rwandan refugees fleeing violence in Burundi, blaming the heavy burden placed by existing refugees in the border districts, especially on the region's fragile natural resources. Continued acceptance of refugees was viewed as 'unsustainable', but the consequence was an abandonment of the protection of human rights. This was not the first time that such an event had occurred, with both Turkey and Honduras in the past citing environmental reasons for why refugees should be excluded from their territories (Ferris, 1993). In a very real sense, in refugee situations, ecology has become politics, and the losers are some of the most vulnerable of the world's population.

The research that is presented in this book was originally motivated by a concern to address this emerging political debate over natural resource management issues in refugee situations. To what extent were refugees placing a burden on the environmental resources of host countries? Were government concerns justified? As the research developed, it became apparent that broader issues were at stake. Not only were refugee movements being seen as a contributory factor in environmental degradation, but so too, 'environmental crises' were argued to be at the root of much forced migration. Humanitarians and environmentalists might be seen as having common cause in controlling both migration and unsustainable use of the environment.

Such a situation is hardly what was envisaged by most of those responsible for developing the terms 'sustainability' and 'sustainable development' over the past two decades. For Redclift (1984: 130), for example, 'the key to redirecting the development and environment discourses lies in the political and economic support given the powerless and the poor. It is an illusion to believe that environmental objectives are other than political, other than redistributive.' But the political nature of ecology is double-edged, and here lies the problem for both policymakers and researchers: is 'sustainable development' at its roots a transformatory or a fundamentally repressive project?

Drawing on field research conducted in a range of countries in Africa, as well as documentary research and interviews with key players about programmes and policies across Asia, Africa and Central America, this book seeks to unravel the political and social issues that underpin natural resource management in the specific case of refugee and post-conflict situations. Such situations provide some unique circumstances in terms of natural resource management, but it is argued here that they can also tell us much about environmental management in more 'stable' conditions of 'development'. There is also much that can be learnt for environmental management in emergencies from the broader development literature.

In conditions of forced displacement, the base required for sustainability, and especially for the kind of transformatory project envisaged by Redclift, may not be readily apparent. But this book argues that a number of points of cross-reference between humanitarian situations and the search for sustainable development are both possible and desirable. At a practical level, measures to monitor and evaluate environmental change, and to address deforestation through the manipulation of supply and demand for woodfuel, for example, show numerous common debates and questions. At a more theoretical level, emerging critical perspectives in 'environmental entitlements' and 'political ecology' are also highly relevant to both fields, and provide pointers to appropriate and worthwhile initial research questions. In both cases, attention to the local specificity of environmental and social change, to the connection between local and global processes, and to the relevance of historical trends in ecology, society and politics, are all important in framing both research and policy.

To a certain extent though, what 'environmentalism' and 'humanitarianism' most have in common, and what marks them out from developmentism and more radical versions of sustainable development, is a shared urgency for those who are professionals or activists in the field. The problems of humanitarian or environmental crises are seen as being of such an order that urgent action needs to be taken to avoid a region, or the world, sliding into catastrophe. However, here too there is great danger. There is a risk that the need to 'do something' can override rational analysis of what exactly the problems are, and how best they might be addressed. Of course, such a criticism might be seen as a bit rich from an academic, given the known tendency of academic writing to debate, to prevaricate, to recommend 'more research' – in short, to do nothing. However, the point of this book is not to suggest that too little is known, or that nothing should be done to address refugee–environment relations, but to take action, aware of the politics of what is being done, and by whom. The most effective counter to this tendency to overeagerness is to ensure that a plurality of voices are heard in defining policies and projects, notably those (usually poor, disenfranchised) rural communities who are supposed to be the main beneficiaries of the action to be taken. The research that underpins this book has sought to do just that, or at least to point in directions that will draw such actors into the debate.

Acknowledgements

The research on which this book is based has been conducted in a variety of contexts. Thanks are due to what was previously the Overseas Development Administration (now the Department for International Development, UK), which funded a project on 'Involuntary Resettlement and Environmental Change', which provided scope both for an initial global review in 1993, and for detailed field research in West Africa in 1994–95. Prior to this, research in Zambia in 1990 was funded by the Economic and Social Research Council (UK), and in 1994 I had the opportunity to work with Care International for a short spell in Tanzania. This latter field level experience in an emergency (or immediately post-emergency) situation in particular taught me much. Subsequently, I have worked as Project Training Co-ordinator on the 'Towards Sustainable Environmental Management Practices in Refugee-Affected Areas' (TSEMPRAA) project for UNHCR, during 1996–98. This provided an opportunity to follow up hypotheses gleaned from work in West Africa, and conduct further research in a much wider context. It also gave me a huge insight into the workings of UNHCR, in a warts-and-all review that it was courageous for UNHCR to implement, and two major donors – the Dutch and US governments – to fund. Whilst numerous individuals in each of these organisations have contributed to the development of ideas contained in this book, neither they, nor the organisations, bear any responsibility for errors or opinions expressed in the final product.

Specific thanks are due to a number of individuals who either collaborated in field research, contributed great ideas (often over a beer), or gave their time to reading draft chapters of the book or earlier working documents. I could not have completed the field research in West Africa without the collaboration of Mohamed Sessay, now of the University of Leeds. He and Ken Wilson, who I had the pleasure of working with in Zambia, are field researchers *par excellence*. I also benefited greatly from collaboration with Thomas Mabwe (Department of African Development Studies, University of Zambia) and Florence Shumba in Zambia, Aminata Niang and Moctar Cissokho (Department of Geography, Cheikh Anta Diop University in Dakar) in Senegal, and Faya Jean Milimouno (Department of Sociology, University of Conakry) in Guinea. Caiti Steele provided valuable research assistance at King's College London in the early phases of writing up, whilst Melissa Leach, James Fairhead, Bill Morgan and Barbara Harrell-Bond have all been a constant source of encouragement as the research developed. Thanks are also due to Andrew Collins, Caroline Dennis, Sanjay Dhiri, Stéphanie Guillaneux, Herman Ketel, Gaim Kibreab, Kibe Muigai, Nontokozo Nabane, Kelly Stevenson, Chris Talbot, and especially Yuki Mori, who provided both ideas and wonderful Japanese food.

As for those who have been asked, or forced, to read draft chapters, I am again sure that the book would not have been finished without them. Thanks go to Chris Barnett, Jean-Yves Bouchardy, Don Funnell, Branwen Gruffydd Jones, Yuji Kimura, Gus le Breton, Francesca Naylor, Matthew Owen, Bernie Ross and Elizabeth Umlas, who read through drafts of part or all of the book and provided me with useful commentaries and reactions. Finally, thanks to my colleagues at Sussex University, and to Martha Walsh, who in varying degrees have inspired me, given me the space to carry out research and to write, and/or simply put up with me.

Abbreviations

CAMPFIRE	Communal Areas Management Programme for Indigenous Resources (Zimbabwe)
CBA	Cost–Benefit Analysis
CBNRM	Community-Based Natural Resource Management
CURE	Co-ordination Unit for the Rehabilitation of the Environment (Malawi)
DFID	Department for International Development (UK), formerly Overseas Development Administration
DHA	Department for Humanitarian Affairs
DRC	Democratic Republic of the Congo
EAP	Environmental Action Plan
ECHO	European Commission Humanitarian Office
EIA	Environmental Impact Assessment
EIS	Environmental Impact Statement
EMP	Environmental Master Plan
ERM	Environmental Resource Management (UK)
ETF	Environmental Task Force
EU	European Union
EWS	Early Warning System
FAO	Food and Agriculture Organisation of the United Nations
FCC	Fuelwood Crisis Consortium (Zimbabwe)
GIS	Geographical Information System
GOM	Government of Malawi
GPS	Global Positioning System
GRID	Global Resource Information Database
GTZ	Deutsche Gesellschaft für Technische Zusammenarbeit (Germany)
ICARA II	Second International Conference on Aid to Refugees in Africa
IDS	Institute for Development Studies (UK)
IFAD	International Fund for Agricultural Development
IGPRA	Income-Generating Project for Refugee-Affected Areas (Pakistan)

IIED	International Institute for Environment and Development (UK)
IOM	International Organisation for Migration
IPCC	Intergovernmental Panel on Climate Change
IUCN	International Union for the Conservation of Nature
LRM	Local Resource Management
LWF	Lutheran World Federation
MSF	Médecins sans Frontières
NDVI	Normalised Difference Vegetation Index
NGO	Non-Governmental Organisation
NRI	Natural Resources Institute (UK)
OECD	Organisation for Economic Co-operation and Development
PIV	Perimètres Irriguées Villagoises
PRA	Participatory Rural Appraisal
PTSS	Programme and Technical Support Section (UNHCR, Geneva)
QIP	Quick Impact Project
RESCUE	Rational Energy Supply, Conservation, Utilisation and Education (Kenya)
REST	Relief Society of Tigray (Ethiopia)
RPG	Refugee Policy Group (US)
RRA	Rapid Rural Appraisal
SAED	Société d'Aménagement et d'Exploitation des Terres du Delta et du Fleuve Sénégal (Senegal)
UN	United Nations
UNDP	United Nations Development Programme
UNEP	United Nations Environment Programme
UNHCR	United Nations High Commissioner for Refugees
UNICEF	United Nations International Children's Emergency Fund
UNRISD	United Nations Research Institute for Social Development
UNRWA	United Nations Relief and Works Agency for Palestine refugees
USAID	United States Agency for International Development
USCR	United States Committee for Refugees
WCED	World Commission on Environment and Development
WFP	World Food Programme
WWF	World Wide Fund for Nature

CHAPTER 1

Why link forced migration and environmental change?

This chapter introduces the topics of forced migration, environmental change and sustainable development. It examines recent increases in the scale and intensity of forced migration, as well as the rising global concern with environmental problems and sustainable development. One reason frequently given in recent years to explain mass population displacement is the growth of environmental problems. Drought and floods, tropical deforestation and sea-level rise are variously blamed for swelling the ranks of refugees world-wide, prompting fears for the security of receiving countries. Such movements of people are also increasingly seen as having environmental consequences. Mass movements of people threaten valuable ecosystems, and risk stretching to breaking point the capacity of particular regions and countries to feed their populations. These two facts alone are seen by many to justify the rising interest in the interrelationships between refugees, environment and development. There is an 'obvious' need for policy interventions to reverse these worrying trends.

However, despite the allure of focusing on what appear to be pressing policy concerns, such reasoning does not provide the primary force behind this book. Instead, this chapter starts the process of critical reflection on how and why forced migration and environmental change have become linked in certain policy discourses, and what the theoretical basis for such focus might be. For example, important initial questions need to be raised: in whose interest is it that refugees should be accused of damaging the environment? Or, in whose interest is it that environmental degradation should be seen as a possible cause of mass displacement? The answers to these questions stress the need to problematise the socio-political context within which discussion of environmental or migration management decisions takes place, and not to accept at face value the terms on which this policy debate is constructed.

Forced migration in the post-Cold War era

Forced migration is a major issue of our times. The number of political refugees – defined under the 1951 Geneva Convention as those forced to

1

leave their homes as a result of 'a well-founded fear of persecution for reasons of race, religion, ethnicity, membership of a particular social group or political opinion' – grew relentlessly during the second half of the 20th century until the mid-1990s. Taking a broader definition of forced migration, 'political' refugees remain only a small proportion of those who are forced to move each year world-wide, with many more uprooted by generalised insecurity, poorly planned development projects and natural disasters. According to the most recent statistics available from the United Nations High Commissioner for Refugees (UNHCR), at the end of 1996 there were just over 13 million officially recognised refugees in the world, with nine million others 'of concern' to the organisation (UNHCR, 1997). Of the total number of refugees, around two-thirds (just under nine million) were living in countries of the Third World. Using a different calculation, the United States Committee for Refugees put the figure at 14.5 million refugees and asylum-seekers in 1996, of whom over 80 per cent were in 'developing' countries (USCR, 1997). These numbers represent a decline over recent years, as a result of significant repatriations, most notably in the latter part of 1996, when over a million and a half refugees were repatriated to Rwanda. None the less, at the end of 1996, it was still possible to identify a total of 32 separate flows[1] of over 50 000 refugees (Figure 1.1), not including Palestinian refugees in the Middle East.[2] If victims of drought and environmental disasters, and those displaced by development projects such as large dams, are included in the total, estimates of the stock of forced migrants run to many tens of millions (*cf.* Westing, 1992).

Of course the existence of a phenomenon, and the interest of donor governments and foundations in it, are not sufficient criteria for the devotion of research time or effort in investigating its causes or consequences. Indeed, one feature shared by the process of forced migration, and that of environmental degradation and change, is that both risk overemphasis because of their striking nature, and, arguably, the threat they pose to the 'established order', to the normalcy of social, economic and political processes. Just as aircraft accidents generate public interest, investigation and state action as a result of their extreme nature, even though this is disproportionate to the number of deaths or injuries actually occurring, so too forced migration and accelerated environmental change have the potential to draw banner headlines and attract public funds, to the detriment of less prominent processes (Cannon, 1995). Some criticism of this overemphasis in the case of environmental issues has recently been voiced by academics (*cf.* Leach and Mearns, 1996) who see at best the creation of a 'discourse' based on popular belief rather than scientific fact, and at worst scaremongering over the future of the planet (and an emphasis on the suitability of scientists to solve it).

However, the point of this book is not to stress issues of forced migration and environmental change because they are 'popular' or striking, but precisely to consider their salience in relation to each other, and in relation to processes of long-term sustainable development. The study of forced migration itself can be justified, for example, at a number of levels. From a purely policy point of view, refugees and other displaced persons represent a 'target group' for international assistance in line with the sometimes

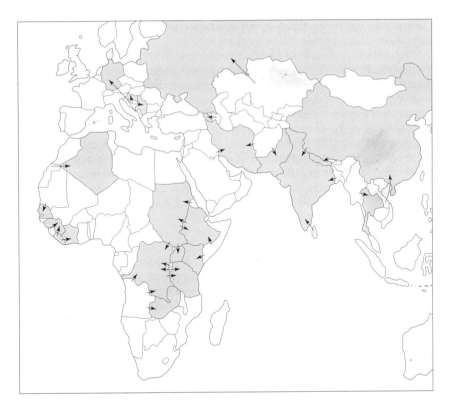

Figure 1.1   Major refugee flows, 1996

dramatic and desperate nature of their plight. As one of the poorest, and certainly most disenfranchised of social groups, refugees attract donor attention, whether from Western governments and international agencies concerned with poverty alleviation, or from non-governmental organisations (NGOs) seeking to reach individuals and families who lie beyond state and social support mechanisms. Given such policy attention, there is a need for independent, academic critique of the actions of governments and agencies promoting the 'well-being' or 'development' of such a marginalised group in society. Independent research is important since in practice, international agencies and NGOs are usually accountable 'upwards' to donors rather than 'downwards' to the beneficiaries of their intervention. Such a critique can act as a check on activities carried out in the name of others, who may be powerless, but could, through well-designed research, become empowered to scrutinise external interventions themselves. Perhaps more obviously, such a critique can also serve to refine and improve the approach of agencies, who after all are carrying out a job that most observers see as highly worthwhile.

However, to be justified as an academic pursuit, the study of forced migration needs to rely on more than simply its policy importance. What distinguishes 'research' in an academic sense from what might be termed 'policy studies' is its concern to look beyond the surface of phenomena at their structure and rationale; and most importantly, to question concepts and typologies not only for their usefulness but also for their theoretical rigour. If we are to study forced migration, we need to be sure – or at the very least have good reason to believe – that the categories we use ('refugee', 'forced migrant', 'returnee', etc.) have some explanatory power. Are these categories that have meaning – either to individuals in terms of their lived experience, or to policy-makers in terms of 'real' differences in behaviour? In other words, are they 'rational abstractions', or merely 'chaotic conceptions' (Sayer, 1982: 71). Are there firm reasons for separating analytically the discussion of forced migration from other kinds of migration, or indeed demographic change more generally? Certainly there is abundant evidence that migration – and specifically forced migration – is important in the shaping of societies and identities. For example, its influence can be seen in the role of slaves, economic migrants and political refugees in the 'making' of American identity over several centuries (Bach 1993) through to the role of forced evictions and war-related relocations in creating Ndebele identity during the liberation struggle in Zimbabwe (Alexander and McGregor, 1997). However, migration 'events' cannot be easily separated analytically from other social and economic changes in either of these examples; nor is the process of migration itself necessarily cited by actors as a salient issue within this broader context.

The contention of this book is that an academic justification for the study of forced migration does exist – and more specifically, that it is academically (or, rather, 'theoretically') interesting to consider relationships between forced migration and environmental change. However, this is far from unproblematic. On the one hand, the very label 'refugee' produces a series of images and socio-political relations that are distinctive and significant for the development of public policy (Zetter, 1988), and this includes the environmental field. At the same time, the process of forced migration throws up situations, categories and constructs that are highly relevant to major social science themes of identity, behaviour, group and community formation and organisation, social transformation, governance, and economic change. People forced to move against their will are equally forced to renegotiate the social, economic and political relationships that tie them to each other, and to the environment in which they live. As an often rapid and dramatic phenomenon, forced migration also frequently throws into stark relief a series of processes that are common in other situations, but are perhaps unobservable, or at least unobserved. A geographical analogy might be the study of coastal landslides, which although quantitatively and qualitatively distinct from 'normal' processes of soil erosion, can none the less tell us something about the latter, as processes and events that might take decades are accomplished by the force of the sea in days or even hours.

Indeed, just as it has been argued that much landscape change is accomplished not so much by slow, imperceptible processes of soil erosion but more by high-intensity, low-frequency events (Thornes and Brunsden, 1977), it is not implausible to suggest that social and political change is more often the result of extreme events (such as wars and associated forced migration) than of, for example, slow, progressive demographic change. A similar example might be the dramatic influence of technological revolutions such as recent advances in global information and telecommunications flows on social processes. However, arguing for such a 'dramatic' world-view is not the purpose of this book.

Environment and sustainable development post-Rio

If forced migration represents a field at risk of over-dramatisation in academic work, then environmental management and sustainable development are surely areas in which academic attention already outstrips the supply of interesting and worthwhile material to discuss. In the last decade, there have been a plethora of books and articles, not to mention videos, CD-ROMs, simulation games, conferences, workshops, NGOs, etc., aimed at informing anyone from the general reader to the specialist of the definition, and importance, of sustainable development. Beyond what is seen by many as the 'standard' definition of sustainable development – the Brundtland Report's 'development which meets the needs of the present, without compromising the ability of future generations to meet their own needs' (WCED, 1987: 43) – there are perhaps hundreds of definitions, each seeking to refine, or in some cases overturn our understanding of what is 'sustainable', what is 'development', and how the two are linked together.

It is certainly not the purpose of this book to attempt to redefine the concept of sustainable development, or to add needlessly to the destruction of forests for paper pulp to make a lengthy case for (or against) sustainable development as a goal of policy. Clearly different definitions of sustainable development are rooted at least in part in different intellectual traditions, and debate in this sense can be seen as simply a part of a much wider theoretical debate. None the less, for the purposes of what follows, it is important to draw attention to two distinctions in sustainable development thinking, which are relevant to the way the concept will be treated in the course of the book. Thus first, a distinction can be drawn between ecologically orientated definitions of sustainable development on the one hand, and human welfare-orientated definitions on the other. The Brundtland definition referred to above is perhaps the archetypal welfare-orientated definition, in which the starting point is 'development', whether this be increasing wealth, the provision of basic needs, or even the securing of human rights. Sustainable development is then about meeting these goals and needs in the long-term without degrading natural or other resources that underpin them. The reason that 'environment' is seen as important according to this tradition is because it is necessary to human welfare, rather than because it is necessarily valued or valuable in itself. In contrast, the ecological approach is perhaps most associated with the World Conservation Strategy, in which primacy is allocated to ecological constraints, and emphasis is then placed on the sustainable utilisation of resources within these constraints (IUCN, 1980). A recent and important elaboration of this position is provided in a benchmark paper by 42 prominent US ecologists (Mangel et al., 1996) which lists seven principles for the conservation of wild living resources, the first of which states 'maintenance of healthy populations of wild living resources in perpetuity is inconsistent with unlimited growth of human consumption of and demand for those resources' (Mangel et al., 1996: 338).

A second, and to a certain extent parallel distinction is between 'strong' and 'weak' sustainability as defined by Pearce et al. (1996). In this case, both concepts are underpinned by the notion that the world possesses a capital stock of resources which determine potential well-being, and which under conditions of 'sustainable development' would not decline in value over time. However, the former allocates to natural resources a special status, such that they cannot be aggregated with human and manufactured resources when assessing overall value. Thus 'strong sustainability' represents the notion that some or all natural resources are non-substitutable by other resources. Absolute protection of particular resources is necessary to ensure that they remain as they are, or at least above certain critical threshold levels in terms of quantity and quality. The failure to maintain a minimum quantity of each resource, and a diverse base of resources in general, is seen as leading to potentially catastrophic results. In contrast, 'weak sustainability' accepts substitution of one natural or human-created resource for another, provided that the overall use value derived for human populations remains the same.

These distinctions are important in the context of this book, since the explicit intention is to draw links between environmental status and a human phenomenon – in this case forced migration, which may result in, or be caused by, environmental change. Taking the case of the arrival of refugees or displaced persons in a country or area, for example, it is reasonable to argue that promotion of human well-being needs to take primacy, given the 'emergency' status of many forced migration situations. The priority for governments and assistance agencies is to ensure the provision of basic necessities to affected populations. In turn, the immediacy of welfare concerns helps to explain why environmental issues have traditionally been seen – and indeed are often still seen – as a side issue in refugee emergencies. Under whatever variant, 'sustainable development' is about conserving natural resources in the medium to long term, whereas refugee assistance is almost by definition concerned with the short term. As such, 'sustainability' is only likely to be taken on board as a priority for refugee assistance agencies if this means 'human welfare' sustainability, rather than longer-term planetary concerns. Similarly, it is the slow undermining of the sustainability of individuals' or households' livelihoods that is generally seen as the most prominent connection between environmental degradation and the production of refugees (Suhrke, 1993). So-called 'environmental refugees' are unlikely to result, at least in the short-term, from threats to global biodiversity.

However, even under a human-welfare orientated 'weak sustainability' approach, over-use and consequent rapid degradation of natural resources *can* be seen to matter, since it is clear that such degradation may undermine human welfare for particular groups in the short term. There is also some interest in focusing on 'strong sustainability' concerns, given that the irreversibility of certain changes in natural resources, uncertainty about the workings of ecological systems, and problems of scale, may all convert relatively small disruption to ecosystems into much larger-scale damage as threshold limits are exceeded. Both provide a policy justification for interest in the impact of refugees on the environment, although from a 'weak' sustainability perspective, conserving resources can be seen as a part of the process of assisting refugees, whereas from a 'strong' sustainability perspective, pursuing 'environmental' goals may conflict with this process.

In developing their argument about 'strong' and 'weak' sustainability, Pearce et al. (1996) focus on the different kind of indicators that can be used in assessing whether a situation or process is sustainable or not, contrasting measures of 'carrying capacity' and 'resilience' associated with the former approach with those of a 'green national income' and 'genuine savings' associated with the latter. Of these, the notion of 'carrying capacity' is perhaps the most well-known, as well as being the one most obviously linked to processes of forced migration which increase local concentrations of population. The risk that high and increasing population densities will exceed the carrying capacity of an area is one that is deeply engrained in popular thought; however, it is also increasingly criticised both from within ecology and beyond, and deserves examination before proceeding in more detail to the links between forced migration and environmental change.

Essentially, there are three major strands of criticism of the concept of 'carrying capacity', which can be loosely described as technological (or economic), institutional (or political), and ecological. Technological arguments were most forcefully stated at a global level in response to the publication in 1972 of *The Limits to Growth* by the 'Club of Rome' (Meadows et al., 1972), which postulated that the world's population had exceeded, or would soon exceed, its carrying capacity; but they can also be seen in the ongoing debate between neo-Malthusian and Boserupian views on the relationship between population and agricultural growth at a regional or local scale. Thus in contrast to the absolute limits postulated by neo-Malthusians, advocates of a technological approach argue that 'carrying capacity' is constantly changing as technological advances increase the productivity of natural and other resources (Anon., 1972). The contribution of Esther Boserup (1965) to this debate is particularly poignant to situations of forced migration, since she argued that it is population pressure itself which provides an economic and/or social imperative to innovate and develop new technologies. For example, rapid population growth experienced in areas hosting refugees might lead to technological change, for example in farming systems. This is a process that has been observed in some refugee situations (see for example Black and Mabwe, 1992), although such innovations can be seen as partly related to the interaction of ideas, rather than growth itself.

7

A second criticism of the principle of 'carrying capacity' comes from a more institutional or political dimension, in which it is postulated that the adequacy of resources to meet human needs depends less on the absolute ratio between population and resources, but more on the distribution of resources, and specifically the way in which social, economic and political systems control this distribution. In an important response to the Club of Rome report, and other literature on the 'population explosion' and its likely impact on natural resources, Harvey (1974) developed the argument that far from being caused by excess population, scarcity of resources is necessary to the capitalist system, since it is this scarcity which creates added value and thus the opportunity to generate profits. At a more local level, it is arguable that relations of power between different social groups will determine not only access to natural resources, but also the overall productivity of natural systems, since mechanisms of exclusion may limit the ability of different social actors to promote productivity improvements. Once again, the marginal status of many displaced populations makes this a very pertinent point in forced migration situations: in effect, such situations may highlight unequal power relations, and the way in which these are translated into resource scarcity through unequal access to natural resources.

Thirdly, criticism of the concept of 'carrying capacity' has come not only from economics and the social and political sciences, but also increasingly from within ecology itself. There is increasing recognition of the dynamic nature of ecosystems, in which the notion of a 'climax' vegetation, towards which particular ecosystems will tend, is replaced by the concept of a dynamic equilibrium between a particular state of vegetation and a series of factors such as rainfall, temperature, humidity, wind, prevalence of fire, human activity, etc. The latter perspective views the natural environment not as a stable 'given' which human interference disrupts, but more as a constantly evolving set of relationships between resources and people (Behnke et al., 1993; Scoones, 1995). Moreover, in this conception, environmental change encompasses both slow gradual processes and relatively catastrophic events, each of which contributes to a new dynamic equilibrium.

Why forced migration and sustainable development?

As hinted at above, at first glance, the relationships between forced migration and sustainable development may not be immediately obvious. After all, those who seek political asylum because they are forcibly uprooted would appear to do so because of processes that are quite the opposite of 'sustainable' or 'developmental'. Whether created by war, armed conflict or political persecution, large-scale refugee movements are generally (and rightly) seen as humanitarian crises – extreme events outside the 'normal' order of things. But the links between the two seemingly dissimilar and incompatible phenomena of forced migration and environmental change go deeper than the fact that both have accelerated in importance over the last decade or so, or that correspondingly, both have received increased international academic and policy attention during that period. Indeed, the above section started to explore some of the reasons why the impact of forced migration on the demand for natural resources might be interesting, in policy terms as well as in illuminating important and current intellectual debates on sustainable development and environmental change.

In practice, potential relationships between war, forced migration and environmental change, and specifically their policy impacts, go much further than the impact of refugees on the environment. (An initial point to make is that the link between war and environmental destruction is well known and relatively well understood, even if the consequences – most notably the fact that large tracts of land in countries such as Bosnia, Angola, Mozambique and Cambodia have been made unusable through the placing of anti-personnel mines – have not been easy to address. This link first received significant academic attention during the Vietnam War (Lacoste, 1973; Wagner, 1974; Westing, 1976) and the anti-Portuguese wars in Mozambique and Angola (Roder, 1973), with the widespread use of such products as napalm and Agent Orange. This had devastating effects both on the flora and fauna, and on the human populations of affected areas. More recent wars have also had significant negative consequences for the environment. Formoli (1995: 68), for example, cites deforestation, damage to irrigation systems, loss of wildlife, and 'uncontested (i.e. neither confirmed nor denied) reports of radioactive waste dumping' by the former Soviet Union during the war in Afghanistan. The civil war in Uganda has involved massive loss of wildlife, whilst another, better publicised example of conflict-related damage was of oil fires and spillages caused during the Gulf War in Kuwait (Roberts, 1992). None the less, despite this recognition, Shaw (1993: 113) bemoans an 'information vacuum' on the environmental impacts of war, noting that *Our Common Future* devotes only a few pages to the dangers of nuclear war (WCED, 1987). The subject was also only marginally discussed at the Rio 'Earth Summit' in 1992 (Bouvier, 1992).)

In general terms, since Rio, there has been a surge of interest in protecting the environment amongst local and international organisations, individuals, and perhaps to a lesser extent, governments. Environmental concern can be seen to have its roots in an increasing recognition in Western societies that the standard of living to which we have become accustomed is unsustainable in the long term – whether in the 'strong' or the 'weak' sense outlined above. The rise of Green parties and movements, although highly contested, has forced environmental and sustainability issues onto the political agenda. Nowhere, perhaps, is this more the case than in development studies and the practice of 'development', although the irony (or hypocrisy) of focusing on the 'destructive' activities of the world's poorest people when so much damage has already been done in the name of 'progress' by the richest nations of the world is inescapable. In many respects, discussion of 'sustainable development' has become coterminous with discussion about meeting the livelihood needs of the world's poorest, and this has come to include populations whose liveli-hoods have been disrupted, and in some cases devastated, by war and forced displacement.

One highly visible consequence of this surge of interest has been the fact that it has become increasingly in the practical interest of a range of both development and humanitarian organisations to take on board environmental concerns. For example, since Rio, it has been incumbent on UN agencies to seek to make their programmes and policies more 'environmentally friendly'. At the same time these organisations have also been forced to deal more and more with situations of conflict and displacement – so-called 'complex political (and environmental) emerg-encies'. Thus the UNHCR has been encouraged, both financially and politically, to examine the environmental consequences of its activities, as well as the broader environmental context in which it operates. A range of NGOs working in refugee situations have sought to become aware of the environmental context of their activities – or perhaps less charitably, they have responded to the necessity to address environmental concerns in order to secure funding for their projects. Whether 'environ-ment' has become a deep-seated concern or just 'trendy', the fact remains that for an increasing number of international donors, environmental activities have become either eminently fundable, or a requirement of any project proposal. In turn, this increasingly applies whether the pro-posed project is 'developmental' or 'humanitarian'.

Development organisations, whose concerns might seem more imme-diately in tune with the concept of 'sustainable development', are also increasingly being forced to consider the implications of war, conflict and mass displacement on their 'normal' developmental and environmental activities. For example, the more obviously 'environmental' organisations within the UN such as the Food and Agriculture Organisation (FAO) and the United Nations Environment Programme (UNEP) are becoming con-cerned with 'complex political emergencies'; whilst the World Bank has recognised some of the problems associated with development-induced displacement – notably displacement of people by large dam projects – in terms of their impact on local environments (Cernea, 1990). As Sorenson (1994: 181) notes, 'large numbers of refugees may have an impact on fragile African environments, and thus pose additional problems for develop-ment'. However, the following section deals first with more detailed ex-planations for increasing interest in the other side of refugee–environment relations, namely the rise of so-called 'environmental refugees'.

Environmental explanations of migration: whose agenda?

In some respects, the link between forced migration and environmental change is most clearly seen in the sense that severe environmental de-gradation might be expected to cause forced migration. The evidence for such a process is reviewed in more detail in Chapter 2, but even without any evidence, it is easy to see why there might be cause for concern – especially amongst governments. As noted above, environmental degradation, whether 'natural' or human-induced, brings with it the risk of undermining people's livelihoods to such an extent that they are forced to migrate.(Myers and Kent (1995: 18) describe environmental refugees as 'persons who no longer gain a secure livelihood in their traditional homelands because of what are primarily environmental factors of unusual scope'.) Whether associated with drought and 'desertification' in the Sahel, an increased frequency of flooding in Bangladesh and other low-lying delta areas around the world, or dramatic events such as earthquakes, or the nuclear meltdown at Chernobyl, large-scale environmentally induced migration poses at best a challenge to planning authorities, and at worst a threat to civil order and the ability of social, political and economic systems to cope. It is easy to see both the origin and the symbolic force of the notion of 'environmental refugees', although the term itself has been vigorously criticised by, amongst others, McGregor (1993) and Kibreab (1994) for being poorly defined and legally meaningless and confusing. It could also be argued that the phenomenon is hardly new, in that environmental factors have clearly been a factor in population movements for centuries.

Before going any further, it is important to note some of the difficulties and the controversy that the term 'environmental refugees' has generated, since they highlight the intensely political nature of its usage. Generally, the term is traced to a UNEP policy paper by El-Hinnawi (1985), although Kibreab (1997) points out that its origin was an International Institute for Environment and Development (IIED) briefing document in November of the previous year. Certainly it has subsequently been the subject of numerous reports emanating from international organisations (IOM/RPG, 1992; Trolldalen et al., 1992) and think-tanks (Jacobson, 1988; Hazarika, 1993; Suhrke, 1993), and amongst others, Norman Myers in particular has been prominent in popularising the term amongst dignitaries ranging from President Clinton to the then United Nations secretary-general, Boutros Boutros-Ghali (Kibreab, 1997). As Kibreab points out, 'prominent international personalities are irrelevant in determining the explanatory or predictive value of a term' (Kibreab, 1997: 21) – but they are important in allowing it to gain currency.

This observation begs the question then of why the term 'environmental refugee' has been so seductive. For Kibreab (1997: 21), the answer lies in the agenda of policy-makers in the North, who wish to further restrict asylum laws and procedures: thus the term was 'invented at least in part to depoliticise the causes of displacement, so enabling states to derogate their obligation to provide asylum'. Since current international law does not require states to provide asylum to those displaced by environmental degradation, argues Kibreab, the notion that many or even most migrants leaving Africa for Europe, or Central America for the USA, are forced to move by environmental factors allows governments to exclude a significant number from asylum. Academics have in turn been complicit in this process by endorsing the term. This is a plausible explanation, given some force by Westing's (1992: 205) observation that the ranks of both recognised and unrecognised refugees 'are being swelled by environmental refugees rather than by political or social refugees'.

However, the notion that 'environmental refugees' have been talked up by northern governments seeking to restrict asylum sits somewhat uneasily with the fact that much of the literature on 'environmental refugees' has in practice argued for an *extension* of asylum law and/or humanitarian assistance to cover those forcibly displaced by environmental degradation, rather than endorsing a differentiation between 'political' and 'environmental' causes as a matter of policy. Thus a report by the World Foundation on Environment and Development and the Norwegian Refugee Council (an arm of the Norwegian government) argued for establishing a system of protection for environmental refugees (Trolldalen et al., 1992: 23), whilst the International Organisation for Migration (IOM) and the US-based Refugee Policy Group (RPG) also concluded that new international instruments were needed to provide assistance and/or protection to a group currently ignored by international policy (IOM/RPG, 1992: 30). Even if the practical impact of literature on 'environmental refugees' has been to endorse northern states' moves to restrict the definition of asylum still further, this does not appear to have been the conscious intention of many of those writing on the subject.

In fact, one of the ironies of writing on environmental refugees has been that whilst purporting to highlight a 'forgotten' category of forced migrant which is ignored by international policy-makers, this literature in practice serves only to differentiate a single cause of migration, which often forms part of a set of reasons why an individual or family may be forced to relocate. As McGregor (1993: 158) argues, 'the use of the term "environmental" can imply a false separation between overlapping and interrelated categories'. But this separation is frequently not made in practice by organisations such as the UNHCR who already use their 'good offices' to provide assistance to a range of groups in 'refugee-like circumstances.' In this sense, then, Kibreab is correct in stating that to focus on 'environmental' causes could lead to the withdrawal of asylum from those who currently receive it – except that the focus here would be much more on large-scale forced migrations inside the developing world (where the UNHCR, for example, has much more room to manoeuvre in influencing which populations should receive protection and assistance, and where states have not traditionally screened individual asylum applicants), rather than on asylum in the North, where rules are already very restrictive.

If academic and policy interest in the notion of environmental refugees is not overtly motivated by a desire to restrict asylum, the question remains as to why so much effort should have been spent in trying to separate environmental causes of migration from other political, economic or social causes, even to the point of trying to rewrite the definition of a refugee in international law. Arguably, the answer lies not in asylum literature or policy at all, but in environmentalist literature, as well as in the field of 'conflict studies'. Thus one of the major proponents of the notion of 'environmental refugees', Norman Myers, comes not from a background in migration or refugees or asylum, but from the science of ecology; in turn, the principal concern of his writing is not migration, but the imminent threat of environmental catastrophe surrounding climate change (Myers, 1993a, b), deforestation and desertification (Myers, 1993c). In an article for the magazine *People and the Planet*, he points out that he 'does *not* assert that the immigrant problem should be perceived as some sort of threat' (Myers, 1993d: 28). None the less, he goes on to suggest that 'without measures of exceptional scope and urgency, Europe may have to accommodate growing numbers of newcomers', and poses an ominous choice: 'either to be more expansive in our attitudes towards neighbouring countries that are also developing countries, or accept that Europe's living space will have to become more expansive to accommodate extra people'. In other words, to do something about the rising tide of environmental refugees also requires governments to do something about the causes of environmental degradation.

This in turn was a point that had not escaped the organisers of a 1994 UN symposium on 'Desertification and Migration' at Almeria, in which the Intergovernmental Negotiating Committee on a Convention to Combat Desertification (INCED) sought to generate northern (i.e. donor) interest in this forgotten convention by highlighting the threat of mass migration to northern countries if nothing were done. Thus the 'Almeria Declaration' produced by the symposium states:

> The number of migrants in the world, already at very high levels, none the less continues to increase by about 3 million each year. Approximately half of these originate in Africa. These increases are largely of rural origin and related to land degradation. It is estimated that over 135 million people may be at risk of being displaced as a consequence of severe desertification. (INCED, 1994: 1)

The driving forces behind this declaration were the representatives of southern rather than northern governments; indeed the northern academics who attended were principally responsible for 'talking down' the figure of populations 'at risk of being displaced' from an initial one billion. It is also interesting to note Kibreab's observation that the term 'environmental refugee' originated with the United Nations Environment Programme – the first, and one of the few UN organisations not to be located in the North, and seen by many as being more firmly aligned to African rather than 'northern' interests within the UN.

Perhaps more important still in pushing the notion of 'environmental refugees' to centre stage have been writers in the field of conflict studies, as attention has shifted away from superpower rivalry as the major cause of conflict and forced migration after the fall of the Berlin Wall and the end of the Cold War. For example, reporting on a major project sponsored by the American Academy of Arts and Sciences and the Peace and Conflict Studies Program of the University of Toronto, Thomas Homer-Dixon (1994) presented three hypotheses on the relationship between environment and conflict: (a) that environmental scarcity leads to simple scarcity conflicts between states; (b) that environmental scarcity causes large population movement, which in turn causes group-identity conflict; and (c) that environmental scarcity causes economic deprivation and disrupts social institutions, leading to 'deprivation' conflicts. Although Homer-Dixon rejected the first hypothesis, the latter two were upheld, focusing for example on Bangladesh and north-east India (Assam) as a case in which millions of environmentally displaced people have contributed to communal conflict.

This theme has been taken up by Suhrke, who draws a distinction between 'environmental migrants', who respond to a combination of 'push–pull' factors (prominent amongst them environmental factors) and 'environmental refugees', suggesting that 'if it is to have a meaning at all, the concept of environmental refugee must refer to especially vulnerable people who are displaced due to extreme environmental degradation' (Suhrke, 1993: 9). This distinction is seen in part as having a temporal aspect, as the slow build-up of environmental degradation is associated with 'environmental migration', to be followed by the reaching of a threshold point at which sudden, absolute and irreversible degradation induces a flow of refugees. However, such a distinction begs a number of questions, not least how a 'refugee' is defined; as McGregor (1993) notes, the legal definition of a refugee – and ultimately the one that guides government and international policy – centres not on the speed of the onset of migration, or primarily on whether it is 'forced', but on the crossing of an international boundary and consequent need for protection that cannot be, or is not, provided by the country of origin. Thus in circumstances where an individual satisfies the criteria for being labelled a 'refugee', the term 'environmental' becomes redundant. In turn, it is unclear that the complex set of factors that lead to 'environmental migration' as defined by Suhrke would suddenly evaporate or crystallise into a single 'environmental' cause at the time people become refugees. Although a distinction could be sustained at the level of proximate causes of flight, this is unhelpful from an academic point of view if it is accepted that the response to forced migration needs to be guided by underlying, rather than simply proximate causes.

Meanwhile, there are perhaps as many typologies of 'environmental migrants' and 'environmental refugees' as there are papers on the subject. Thus El-Hinnawi (1985) and Jacobson (1988) started with three subcategories of environmental refugee, namely temporary displacement due to temporary environmental stress; permanent displacement due to permanent environmental change; and temporary or permanent displacement due to progressive degradation of the resource base. In contrast, IOM/RPG (1992) draw distinctions between emergency versus slow-onset movements; temporary, extended and permanent movements; and internal and international movements, whilst accepting the point above that the migrant–refugee distinction hinges on the issue of protection. Suhrke (1993) divides her discussion into migration stimulated by deforestation, rising sea levels, desertification and drought, land degradation, and water and air degradation, before proceeding to identify environmental pressure points at which the combination of such factors establishes a susceptibility towards environmental migration. Trolldalen et al. (1992) distinguish between 'refugees' from natural disasters; degradation of land resources; involuntary resettlement; industrial accidents; the aftermath of war; and climatic changes.

Most recently, the ball has returned to the IOM (1996), which uses a sixfold division similar to that of Trolldalen et al., but which draws an overall distinction between 'natural' and 'human-made' causes. Taken as a whole, the impression is one in which lists of factors have overcome theoretical rigour.

Environmental consequences of migration: protection and burden-sharing

If an intensely political agenda can be perceived beneath the surface of calls for more attention to 'environmental refugees', the same is no less the case for the current growing interest in the environmental consequences of forced migration. And yet, once again, at first sight, the nature of concerns is clear; as Sorenson (1994: 181) notes: 'Typically, areas in the immediate vicinity of refugee camps are stripped of vegetation cover because wood is needed for cooking and shelter. This alters soil and water balances and leads to erosion, soil depletion and decreased productivity.' Whilst clearly a generalisation, Sorensen's comment has a resonant tone for anyone who has visited a refugee camp. Vegetation loss is visible, tangible evidence of the presence of refugees, a clear reminder of abrupt change that is frequently brought not only to environmental systems, but also often to social and economic systems as well. Moreover, such change is not dependent on the cause of forced migration, applying equally to forced migration internal or external to national borders, whether caused by political, economic or social upheaval.

There are a number of reasons why environmental degradation is becoming an issue of greater concern not only to governments of countries hosting refugees, but also to the 'international community' more generally, including agencies working with refugee populations and donor governments. First, from a purely practical point of view, it is very often the case that displaced populations are forced to rely on natural resources available to them locally in order to sustain themselves in the short term. However rapid the international response, there is nearly always a period of time between the first arrival of refugees in a country or area and the arrival of assistance in the form of food aid, blankets, medical supplies and so on; and even once (or if!) these 'basic necessities' are provided, there are still commodities which are not generally supplied by external donors, notably fuel for cooking and heating (see Chapter 4). In this sense, concern for the environment, and specifically for the sustainability of natural resource use, is of direct concern to assistance agencies in that it constitutes a major part of the 'livelihood' of the refugees themselves. Failure to protect this 'livelihood environment' will at best make it more difficult for refugees to find essential commodities, especially if they are in exile for a long period, and at worst may contribute to malnutrition, disease and enhanced poverty.

There is also an important additional side to the relationship between people and their 'livelihood environments' in refugee-hosting areas which has deeper ramifications for the whole system of asylum and refugee protection itself. Put simply, it is not only refugees who rely, to a certain extent, on their immediate surroundings for basic items such as wood, forest products, grazing land and the like; the local populations of the refugee-hosting area are also likely to be reliant on the same resources, and have a stake in their sustainability in the longer term. Unsustainable use of natural resources (or indeed of infrastructure, such as schools, roads and clinics) threatens the livelihood environments not only of refugees but also of local host populations. The result may be an undermining of the willingness of the latter to accept the presence of the former, and may in extreme cases lead to conflict over natural resources between the two populations.

Environmental concerns in refugee-hosting areas are not confined, meanwhile, to immediate impacts on the livelihood environments of affected populations, and it is here that a more significant justification for the concern and involvement of international agencies and donor governments can perhaps be found. In particular, a principal environmental concern world-wide in recent years has been the issue of loss of biodiversity, with international responses ranging from the negotiation of the Convention on Biological Diversity (1993) to the elaboration of management plans and conservation agreements to protect areas of special importance, such as forests, wetlands or other habitats that support particularly diverse or rare species of flora or fauna. In this context, it is easy to see how sudden influxes of migrants can be seen as a major threat to plans or agreements which have been painstakingly worked out with local inhabitants of such areas. In essence, the immediate livelihood of people seeking to survive comes into conflict with longer-term and more global concerns of protection of endangered species and 'saving the planet'. Of course, a simple answer would be to avoid refugee settlement in areas of conservation importance, but unsurprisingly, the geography of war and mass flight has tended to pay little attention to the geography of protected areas. Indeed, given that many protected areas already represent contested spaces, there seems more likely to be spatial correlation than spatial avoidance.

Evidence of unwillingness to accept the presence of refugees as a result of the 'environmental damage' they cause can already be observed in the development of state policy towards refugees around the world. As noted in the preface to this volume, the governments of Honduras, Turkey and Tanzania have all cited environmental problems as justification for closing their borders with neighbouring states. However, the politics of such decisions are not a straightforward response to the cost of environmental damage – not least because the evidence for real long-term damage caused by refugees (as opposed to reasons why it might be a problem) is still sparse, as is discussed in Chapter 2. Rather, there are a number of reasons why it might be politically convenient for both governments and international agencies to 'blame' refugees for environmental damage, not least because they are a soft target with few rights or opportunities to defend themselves and their actions. For governments, there is certain convenience in being able to cite environmental destruction as a legitimate reason for stricter controls on asylum – or for rejecting unwanted refugees outright. In doing so, African states at least are simply following the lead of Europe, for example, in placing restrictions on asylum; but they are able to deflect the (largely hypocritical) criticism of European nations and agencies by couching their policy in environmentalist language, which in many respects is more resonant in Europe than the language of human rights. This is a problem that is compounded by stereotypical representation by the Western media of large-scale population displacements. Violent political conflicts in Africa are explained in terms of 'tribal conflict', 'barbarism' or 'apocalypse', implying that they are beyond the comprehension of the civilised West. As such, it is hardly surprising if saving the rain forest, or protecting gorillas, proves more important to European public opinion than the plight of refugees.

However, lying behind the convenience of 'blaming the victim' – in this case refugees – there is a more serious point to be made, and indeed, a more serious concern of host governments. For in so far as hosting large refugee populations does represent a 'burden' on host governments and societies, this burden falls disproportionately on developing countries. Yet despite this global imbalance in the offering of asylum, there is little talk of 'burden-sharing' on the part of the richer countries of the world, or indeed even recognition of the cost to poor host countries of providing asylum (although there are clearly both costs and benefits, and as Chambers (1986) rightly pointed out, these are likely to accrue to different groups even within refugee-hosting areas). Rather, the signals on asylum put out by richer countries are quite the opposite: a clamping down on third country resettlement; fines imposed on carriers who bring in asylum-seekers; individual assessment of asylum claims and harsh implementation of the letter rather than the spirit of the Geneva Convention; and deportation of those who do not meet the criteria, often not directly to the country they had fled from, but to the (usually poor) third country through which they have transited.

Refugees as 'exceptional resource degraders'

Earlier in this chapter, the question was raised as to whether forced migration brings with it social processes or actions that are distinct from those which might be observed in the 'normal' course of events. The existence of such a distinction could be seen as justification for treating forced migrants – or refugees – as a separate analytical category. One such argument that merits discussion is the contention that by the nature of their circumstances, forced migrants are more likely to contribute to environmental degradation than non-migrant populations. The notion that refugees are 'exceptional resource degraders', first raised by Leach (1992) in the context of settlement of Liberian refugees in Sierra Leone, is based on the principle that refugees have no long-term stake in the sustainability of the natural environment. To this basic position, various authors have added further reasons why refugees may not use resources in a sustainable way, independent of their numbers or the quality of the land on which they are settled. Leach herself suggests the lack of long-term commitment to the host area may combine with refugees' poverty to lead to short-term resource exploitation. Echoing this, Jacobsen (1994: 12) quotes the argument of Myers (1993e) that 'in several respects [people living in absolute poverty, such as the displaced] appear to cause as much environmental damage as the rest of the developing world combined'. Added to these factors, Jacobsen suggests that refugees may be unfamiliar with the host environments in which they are living, whilst a process of 'commoditisation' may occur in which 'free' local resources acquire a monetary value and so become overexploited because of their potential to earn income. This process, she argues, will depend partly on the nature of refugee–host relations, and of international assistance.

Building on this theme, Ketel (1994a) notes that the trauma of war and resettlement, and the lack of ownership of land, may reinforce refugees' lack of incentive to observe principles of sustainable resource management. Hoerz (1995a) adds that refugees are not governed by traditional common property resource management systems, whilst there may be a breakdown of social authority within the refugee population. He also suggests that insecurity around refugee camps and settlements may reduce the radius of exploitation, whilst 'refugees transport essential goods like water, firewood and thatching material mostly on foot' (Hoerz, 1995a: 29), although it is difficult to see how they are different to many non-displaced African rural populations in this respect. Leach's study in Sierra Leone found some evidence of refugees acting in unsustainable ways, with an example cited of 'occasional instances of stranger refugees cutting down whole palm (trees) in order to eat the hearts or obtain the fruit quickly and easily' (Leach, 1992: 44). She also questions whether the common pattern of refugee agriculture – planting cassava on fields after the rice harvest and shortening the length of fallows – might have led to a reduction of soil fertility, although there was no firm evidence of fertility decline, and local farmers did not recognise this as a problem, and indeed were beginning to farm 'second year' cassava plots themselves.

However, although the notion of refugees as 'exceptional resource degraders' is eminently plausible, in many respects it is a deeply flawed concept. First, it is surely not possible to generalise about the human behaviour of such a large group of people in such a complex relationship as that between people and the environment. Politically, there is an inevitable undertone of 'blaming the refugee' in the suggestion that refugees unreasonably ignore sound management practices in their use of natural resources. Empirically, the concept barely holds up to scrutiny. Thus despite introducing the phrase for discussion, Leach's overall conclusion in Sierra Leone was that environmentally destructive practices were rare, and that in general the refugees 'successfully acquired food, and gained access to the natural resources they needed for survival without triggering dramatic environmental decline' (Leach, 1992: 46). Objections to the notion can also be sustained at a theoretical level. Perhaps the most comprehensive rebuttal of the 'exceptional resource degrader' thesis is provided by Kibreab (1997), who develops his argument on the themes of the role of poverty, the supposed lack of incentives for refugees, uncertainty of refugee situations, unfamiliarity with host environments, and spatial segregation of refugees and their consequent isolation from traditional resource management systems. Whilst admitting that a temporary attitude towards resource management on the part of refugees might limit strategies that would promote sustainability, he argues that such an attitude is more characteristic of governments and assistance agencies, rather than of refugees themselves.

Taking first the question of poverty: whilst it is largely true that poor people live in the worst environments, this is far from saying that poverty creates or encourages environmental degradation. For example, Myers (1993e) is cited above, as well as by both Kibreab (1994) and Jacobsen (1994), as supporting the view that poor people overuse natural resources. However, he appears at the same time to contradict this argument, talking instead of an 'affluence multiplier' (Myers, 1993e: 206), in which environmental degradation increases as wealth increases, and presumably as poverty decreases. In turn, Kibreab (1997) directly asks the question, 'does poverty (of refugees) cause environmental degradation?' concluding from his own survey evidence in eastern Sudan that it does not. Instead, he identifies mechanised commercial rain-fed crop production, clearing of vegetation for tractorisation, and commercial firewood and charcoal production for cities as the main causes of resource depletion, whilst arguing theoretically that poverty motivates more prudent and conservative use of resources, since the consequences of careless use might be calamitous for the refugees. Such an argument touches on the much wider question of the relationship between poverty and environmental degradation, a relationship which at best is not straightforward (Leach and Mearns, 1991).

On the issue of the lack of incentives for refugees to conserve the land, Kibreab accepts that there are links between lack of tenurial security and willingness of refugees to undertake conservation activities, but argues rightly that refugees are not the only groups in Sudan to lack tenurial security. In turn, it is important to look at the roots of this insecurity, which lie not only in eastern Sudan but also in much of Africa in the introduction of new laws on the ownership of land, and particularly the nationalisation of many common property resources. In the Sudan, Kibreab argues this has led both to 'land grabs' by wealthy élites, and to turning common resources regulated by communities into *de facto* open access resources, for which there are no effective controls over use. Kibreab also rejects the notion that refugees use natural resources unwisely because of uncertainty over the length of their stay, or because of unfamiliarity with local environments. On the former, he points out that often, refugees' room for manoeuvre in resource management is highly limited, such that they are naturally risk-averse. On the latter point, he notes that many, if not most refugees actually move over relatively short distances, so that even though they have crossed a national border, they are still highly familiar with both the environments and resource management systems in their host areas. Although in Sudan Kibreab found that refugees were not using the skills and knowledge they derived from their home area in conservation techniques, he argues that neither did local farmers use such techniques, partly because in some areas erosion was not a serious problem, and partly because the necessary 'equipment', in this case stones to construct terrace walls, was not available locally. In any case, many Eritrean refugees in Sudan had shifted from subsistence agriculture in their country of origin to commoditised, mechanised agriculture in their host area.

Finally, Kibreab examines the suggestion that segregation of refugees in sites separate from local communities implies isolation from local traditional resource management systems, and thus higher levels of environmental degradation, but points out that such systems are often already in a state of decay due to state policies such as nationalisation of land noted above. Whilst segregation is seen as problematic, in the Sudan this resulted more from the confinement of refugees, who were thereby prevented from adopting more flexible land management systems than from separation from local rules. However, whilst accepting arguments about the dangers of 'confinement' of refugees, which are developed further in Chapter 5 for the specific case of Rwandans in Tanzania, some caution must be expressed in assuming that indigenous management practices are unable to regulate resource use effectively, whether this is due to state interference or dynamics introduced by the refugees' presence. This issue of institutional mechanisms for resource management is considered in more detail below, and again in Chapters 6 and 7.

The structure of the book

This chapter has outlined various reasons why relationships between refugees, environment and development have come to the fore in a number of policy contexts in recent years. Whilst in one sense, there are persuasive reasons why environmental degradation should be considered as one major cause of forced population displacement, it has been argued that there is also an essentially political agenda which has stimulated interest in isolating such causes. Similarly, for those dealing with refugees, and especially for host governments, there is a political incentive to 'talk up' concern over the environmental damage associated with refugee flows in order to mobilise funds to deal with the problems encountered. However, as the discussion above highlights, separating out refugees as causing 'exceptional' environmental degradation is theoretically questionable, and there is a need to examine the more complex interrelationships between forced migration and environmental change, rather than focusing on simple cause and effect.

The chapters which follow address these questions in a variety of ecological and political contexts. In Chapter 2, a first task is to address the question of how much evidence really exists for large-scale environmental damage being caused by refugees, or of large-scale forced migration as a result of environmental degradation. Although in both cases, some potential linkages can be observed, neither is easily demonstrable on the basis of existing measurement and scientific evidence, and indeed, in some cases, pointers are orientated in the opposite direction. In areas that are labour-scarce, rather than resource-scarce, for example, the presence of refugees may represent a potential labour force for the implementation of 'environmental conservation' rather than environmental damage. The conclusion that there is a lack of evidence of the links between environmental degradation and forced migration leads in Chapter 3 to a discussion of the kind of evidence that might be sought, and the methodologies available to conduct such an analysis. Available methods include the use of remote sensing and geographical information system (GIS) technology, environmental assessments, participatory environmental appraisal, and the process of developing 'environmental action plans' in refugee emergencies, each of which is considered in some detail.

In Chapter 4, attention then shifts to a range of practical environmental interventions that have taken place in recent years in a range of areas that have been affected by large-scale flows of refugees. Drawing on evidence from Africa, Asia and Central America, the chapter discusses interventions including the supply of energy to refugee camps, the promotion of fuel-efficient stoves and alternative cooking technologies, and woodland management and reforestation schemes. This chapter also examines the potential for an approach based on 'community-based natural resource management' – now increasingly popular in developmental situations – as a strategy for environmental management in refugee-affected areas, discussing the differences between humanitarian emergencies, refugee situations more generally, and areas in which the priority is simply 'development'.

Chapters 5 and 6 then examine two case studies of environmental management in refugee-affected areas, respectively in the recent humanitarian emergency in the Great Lakes region of Africa, and in two longer-term refugee situations in West Africa. In the context of the exodus of Rwandans to neighbouring Tanzania and Zaire from 1994 to 1996, Chapter 5 examines issues surrounding site selection, environmental impact assessment, and the development of 'environmental programmes' as part of the relief effort. In contrast, in the case of Guinea and Senegal in West Africa, where large numbers of refugees have been present since 1989–90, the focus is more on local level strategies that have been adopted by refugees and local communities to ensure access to natural resources – and the outcome of these strategies for environmental degradation. These case studies begin to question the value of developing large-scale and specific environmental interventions in situations of forced displacement, and highlight the significance of local context and arrangements for natural resource outcomes whether or not the 'international community' becomes directly concerned with natural resource and 'sustainability' issues.

In Chapter 7, the focus turns to the repatriation of refugees. On the one hand, such repatriation can be seen in part as linked to either actual or (more likely) perceived environmental damage caused by refugees' presence in host areas, although it is argued that this is associated more with a range of political considerations. Meanwhile, on their return to their 'home' countries, refugees find themselves once again in a country where they are citizens rather than 'strangers', and where their participation in political and economic structures which affect environmental management may not be as problematic. Drawing on case studies from a range of countries, the respective roles of international agencies and local institutions is again called into question. This is broadened into a theoretical discussion of the 'extended environmental entitlements' approach to environmental change and sustainable livelihoods, which, it is argued, is particularly pertinent not only to post-conflict situations, but also to conditions of mass displacement more generally.

Finally, in Chapter 8, the book returns to the question of the interaction between forced migration and the environment, through an examination of the obstacles to developing a longer-term, 'sustainability'-orientated view in refugee situations. It is argued that rather than seeking a 'blueprint' approach to environmental issues, what is required in both research and policy is a more flexible, place-specific and yet theoretically informed approach that is aware of both political and historical context. In this endeavour, the language of 'crisis' in dealing with both humanitarian and environmental issues can be seen as unhelpful in unravelling the complex linkages between refugees, environment and development. There is an urgent need to start with understanding, before moving to action. Promoting this understanding is a suitable task for academic research, as well as a crucial step for sound policy.

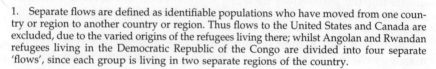

1.   Separate flows are defined as identifiable populations who have moved from one country or region to another country or region. Thus flows to the United States and Canada are
excluded, due to the varied origins of the refugees living there; whilst Angolan and Rwandan
refugees living in the Democratic Republic of the Congo are divided into four separate
'flows', since each group is living in two separate regions of the country.

2.   Refugees in the Middle East are excluded from statistics provided by the UNHCR,
since they fall under the mandate of a separate UN organisation, the United Nations Relief
and Works Agency for Palestinian refugees (UNRWA).

CHAPTER 2

Forced migration and environmental change: the evidence

This chapter reviews a wide range of published and unpublished documents in search of evidence of environmental causes of forced migration, and of the environmental impact of refugees. Although host governments, international organisations and some scholars and 'experts' have been keen to argue that negative environmental changes are occurring in refugee-hosting areas, and/or that the world is filling up with 'environmental refugees', there is surprisingly little scientific evidence to back up either claim. Meanwhile, where such evidence does exist, it is often methodologically flawed. For example, where evidence of 'environmental refugees' is presented, it is often divorced from the social, political and economic context in which assessment of causality might be meaningful. Similarly, where detailed empirical investigation has taken place of refugees' environmental impacts, the wider context is again often ignored.

Environmental change and environmental refugees: the evidence

In Chapter 1, it was noted how estimates of the number of 'environmental refugees', or 'environmental migrants', vary widely, as do definitions and typologies of such flows. However, whatever the precise number and definition, a common feature of the literature is to talk of 'millions' of displaced people, and their dramatic impact on host regions, such that regional security is threatened. The image is one of misuse or overuse of the environment leading to progressive decline in the resource base, and possibly contributing to further dramatic (and unintended) environmental collapse. Environmentalists and conflict specialists see common cause in talk of 'environmental refugees'; even if the linkages between environment, conflict and refugees remain to be proven. It is the purpose of the first part of this chapter to examine the evidence for such linkages.

At first glance, the data available on environmental refugees appears quite impressive. A number of areas of the world are cited by a range of authors as being affected by environmentally induced migration, ranging across Asia, Africa and Latin America. In *Ultimate Security*, Myers (1993b: 189) starts a chapter on the potential for displacement due to sea-level rise (see below) with concern about the plight of Haitian boat people, 'abandoning their homelands in part because their country has become an environmental basket case'. Homer-Dixon (1994: 22) draws, amongst other examples, on the evidence from southern Asia, where the piecing together of demographic information and experts' estimates leads him to conclude that Bangladeshi migrants 'have expanded the population of neighbouring areas of India by 12 to 17 million' over the last forty years, whilst 'the population of the state of Assam has been boosted by at least seven million'. El-Hinnawi (1985) and Jacobson (1988) cite additional examples of environmental refugees from across the Sahel and the Horn of Africa, as well as the Soviet Union and the United States.

However, despite the breadth of examples provided in the literature, detailed argument about why particular situations justify the title 'environmental refugees' is much more scarce, whilst precise statistics are often of questionable accuracy or are simply non-existent. Taking first Homer-Dixon's example of migration from Bangladesh to India, caused by 'environmental scarcity', it is something of a surprise to find that, even in his own article, a number of other explanations for migration vie with that of environmental degradation. Thus migration, we are told, is also associated with rules on land inheritance, the system of water management in Bangladesh, the standard of living in India, and the encouragement of migration by some Indian politicians eager to gain new voters. At the same time, the source of Homer-Dixon's demographic information (Hazarika, 1993) casts some doubt on the statistics too – in this case, between 12 and 24 million such migrants. For example, Hazarika's estimate of migration to Assam comes not from any direct measure, but from a comparison of the 1951 and 1991 Assam census figures, adjusted for the population growth rate in 1951, which shows a notional excess population. Yet this increase could be accounted for in a number of ways, with likely candidates including a rise in the population growth rate after 1951; under-counting in 1951 (eminently plausible in a remote region); or over-counting in 1991 (also eminently plausible given the link between population size and allocation of government resources).

Other accounts of 'environmental refugees' fair little better. For example, a special issue of the Canadian journal *Refuge* on the subject in 1992 provides scant evidence that the phenomenon is widespread. For the case of cyclones and flooding in Bangladesh, Islam (1992) is able to quantify deaths, but not migrants, whilst the two examples given of environmental refugees are less than convincing. One is of rural–urban migrants, hardly a phenomenon peculiar to Bangladesh (or even the present century); whilst the other is of groups resettled in the Chittagong Hill Tracts. In this case, as Islam admits, migration owes as much to government encouragement of settlement in a remote region (in order to combat counter-insurgency problems) as it does to the environmentally induced destitution of the migrants themselves. Otunnu (1992) does little better for Africa, citing Timberlake (1988) as evidence for a 'substantial' number of environmental refugees, but then failing to describe any concrete examples of such refugees. A contribution by a Malaysian NGO, Sahabat Alam Malaysia (1992), speaks of 'the true environmental refugees of our region', but then goes on to describe the Orang Asli tribal peoples, whose lifestyle has been undermined by tourism and development projects. Much as these changes might be regrettable, and indeed opposed by the Orang Asli, little evidence is provided of physical displacement, and those who are mentioned – around 3000 forced to move by a dam project – are perhaps better described as 'development-induced displacees'. The same could be said of contributions to the same edition on people displaced by forest plantation in Thailand (Hubbell and Rajesh, 1992), by the Three Gorges dam in China (Whitney, 1992), and in a slightly different vein, by defolia-tion campaigns in the Vietnam War (Glassman, 1992). It is not that in each of these cases the environment was not radically changed; indeed it was. The point is that in each case, the environment was deliberately changed in pursuit of another goal – such that the 'primacy' of environ-mental factors, integral to most definitions of environmental refugees, is hardly clear.

A search through the references cited in another recent review paper by Ramlogan (1996) reveals little concrete evidence either. Thus for migra-tion from 'natural disasters', we are pointed again to work by Hazarika (1993), as well as to Mattson and Rapp (1991), Sanders (1990–91) and an unspecified report from the US embassy in Addis Ababa. Yet Mattson and Rapp are respectively a climatologist and a geomorphologist, who merely state that 'refugee migration is linked to drought and famine' rather than demonstrating the linkage; Sanders indiscriminately describes as 'environmental refugees' some 4.1 million rural–urban migrants in north-east Brazil in the 1960s and a further 4.6 million in the 1970s, even though he admits that many areas not affected by drought also lost popu-lation as a result of poverty; whilst for Ethiopia, Ramlogan simply repeats an observation originally cited by Jacobson (1988) that one million people 'were about to move because of famine conditions' – without actually saying whether they did (Ramlogan, 1996: 83). Similar difficulties emerge for the effects of 'long-term environmental degradation', where Ramlogan curiously cites the 'Black Triangle' of the Czech Republic, Poland and south-east Germany as a region of out-migration due to pollution (a judge-ment that might be questioned by German authorities currently trying to deal with in-migration!). On the aftermath of war, Ramlogan makes the extraordinary assertion that the failure of Afghan refugees to return to their country from Pakistan is due to poor land productivity and the number of land mines – when surely the continuing conflict in their country of origin, and the largely favourable economic conditions they have experienced in exile, might have more to do with their decision?

Such problems strike to the core of the literature on environmental refugees, and nowhere more so than in the generation of statistics on its prevalence. Perhaps the most quoted statistic is that of Jacobson (1988), whose estimate of 10 million environmental refugees is repeated over and over, but without independent verification of its accuracy. Myers (1996) goes much higher, stating that at present there are at least 25 million refugees, but perhaps advisedly, El-Hinnawi's (1985) earlier paper for the UNEP avoided the issue altogether, instead choosing to present figures of people killed, or 'affected' by various types of environmental problem. In turn, the generation of statistics is critically dependent on the definition of 'environmental refugees', a process which might well be seen as impossible given the multiple and overlapping causes of migration discussed in Chapter 1. In so far as distinctions between causes can be drawn, the following sections consider three different types of 'environmental migration' and the evidence that has been put forward for the existence of these phenomena. It is evidence that is far from convincing.

A 'myth' extended: desertification-induced displacement?

Out of the range of environmental migration 'types' cited above, perhaps one of the most pervasive in terms of popular recognition at least is that of the poverty-stricken (and usually African) farmer who is finally forced to leave the land because of drought, progressive impoverishment of the soil, and ultimately famine. The term 'desertification', conjured up in the 1970s to evoke the relentless onward march of the desert, but with its origins in colonial concern about mismanagement of the environment (Swift, 1996), evokes too the flight of humans towards less hostile lands, or more likely to 'refugee' camps. Particularly in the Sahel, which received considerable media attention as a result of the droughts of the mid-1970s and mid-1980s, but also in other 'marginal' semi-arid areas across Central America, Asia and even southern Europe, desertification-induced migration epitomises the threat supposedly posed to industrialised societies by an army of the poor and starving on the move. As Jacobson (1988: 6) put it: 'desertification . . . has irreparably damaged millions of hectares of once productive land and made refugees out of millions of sub-Saharan African farmers. Migration is the signal that land degradation has reached its sorry end.' However, the evidence for 'desertification' causing migration in any straightforward way is somewhat limited. First, it is important to note that the concept of 'desertification' itself has come under fire in recent years, particularly as availability of satellite images of the region has improved. Thus work by Dregne and Tucker (1988) and Tucker et al. (1991) has shown a highly elastic response of vegetation cover (using the 'normalised difference vegetation index', or NDVI, as a surrogate) to growing season rainfall, with the 'desert margin' in the Sahel fluctuating from year to year as a result. Williams and Balling (1996: 50) question as 'equivocal' evidence that human activities have changed climatic patterns through influencing surface albedo, surface roughness, plant cover and soil moisture. Mortimore (1989) has noted that management practices that were thought to contribute to land degradation need to be placed in historical context. Overall, there is increasing talk of the 'myth' of desertification (Helldén, 1991; Thomas and Middleton, 1994; Swift, 1996).

If one accepts the argument that desertification itself is largely a myth, then it is not, perhaps, too great a step to suggest that desertification-induced migration is a myth too. None the less, even if there is no secular trend of declining vegetation cover and land productivity in the Sahel, and vegetation recovers as rainfall increases, it is possible that relatively permanent migration might result from a temporary decline in the productivity of agricultural and grazing land during drought periods. But what is required is evidence that this is the case. For migrants to be termed 'environmental refugees', it seems reasonable that (albeit temporary) environmental decline should represent the main (if not the only) reason for their flight. In practice, such evidence is hardly forthcoming. For example, in a review of desertification-induced migration worldwide, Schwartz and Notini (1995) cite examples from Mexico, Haiti and the Sahel, as well as the cases of north-east Brazil and north-west India discussed above. But in the case of Mexico, after a review of general environmental problems in the country, there is only a brief discussion of an attempt to correlate statistically areas of emigration with areas of 'aridity', in which the central region is identified as an area that is both arid and suffers high out-migration. Yet the authors admit that not all arid areas are 'degraded', that not all migration from these areas is necessarily the result of desertification. The rather lame conclusion is that 'our discussion with experts, research, and analysis of the relevant statistics data *will likely confirm* that desertification is a factor contributing to migration from this region' (Schwartz and Notini, 1995: 82; my italics). Elsewhere, and reliant on studies funded by the Universities Field Staff International and published by the National Heritage Institute, the evidence is little more convincing: for example, in Haiti, it is stated that deforestation and soil erosion are severe problems facing the country, but no clear link with migration is demonstrated. Indeed, for the case of emigration, it is stated that 'it is evident that most of Haiti's emigrants in recent years have been *political and economic* refugees' (Schwartz and Notini, 1995: 88; my italics), influenced only 'to some extent' by environmental deterioration.

In other work too, the evidence that is presented for migration as a result of drought and desertification is generally only the existence of migration from regions that are prone to such processes. A causal link to drought is seldom established, whilst in some cases, not even the existence of 'excess' migration is demonstrated. Thus Jacobson (1988) cites a number of Sahelian states in which rural–urban or north–south migration occurred during the drought period of the mid-1980s, or in which significant populations became dependent on food aid, and all of this is taken as *prima facie* evidence that these groups have been forced from desert margins because of declining rainfall. In Côte d'Ivoire, foreign immigrants are thought to make up 20–25 per cent of the total population (Russell et al., 1990), whilst there is significant internal migration within the country itself. Both are widely interpreted as resulting from deteriorating environmental conditions further north (*cf.* Wijkman and Timberlake, 1984).

However, within the Sahel, and indeed in other semi-arid regions, there is a tradition of migration that extends back over decades, and often centuries, and which ranges from nomadic pastoralism to long-distance trade, as well as the permanent relocation of individuals and families. In turn, these migrations have their own dynamics, rooted certainly in the difficult environmental conditions of the region, and the need to diversify income-earning opportunities, but not necessarily related to a decline in those conditions (Cordell et al., 1996). For example, there is now an increasing body of literature that has examined migration, both internationally from and locally within the Sahel region, which suggests that a simple link between poverty, environmental degradation and migration is hard to sustain. In the Senegal River valley, one of many source areas for migration to Côte d'Ivoire, Lericollais (1989) notes that migration has long reflected a household strategy to cope with environmental risks, which, although severe, are not necessarily regarded as worsening. Studies have identified how migration plays a cultural role in the transition to manhood, as well as being economically linked to the generation of sufficient revenue to buy livestock (USAID, 1990; Velenchik, 1993). Factors such as the decline of markets for traditional cash crops (which include gum arabic and cotton), the development of Senegal's groundnut basin, and subsequent mechanisation of agriculture in the delta provide additional and more recent motivations to move out of the middle and upper parts of the Senegal River valley (see Adams, 1977). Moreover, such conclusions are not limited to the western Sahel but can be extended across the continent. As David (1995: 18) notes from an empirical study based in Senegal, Mali, Burkina Faso and Sudan:

> Migration does not necessarily signify a rejection of a rural livelihood. Rather, it demonstrates that the survival strategies of rural Sahelians are not only rooted in their immediate vicinity, but are also linked into economies in other rural and urban locations. It is precisely this inter-linkage which supports rural communities and helps them to survive in such climatically unstable environments.

The picture is one of migration as an essential part of the economic and social structure of the region, rather than a response to environmental decline – a picture reinforced by numerous other studies that have confirmed the critical role of migrant remittances in household and regional economies (Condé and Diagne, 1986; Horowitz, 1990).

The situation appears similar in other semi-arid regions of the world allegedly prone to desertification and related migration. For example, Glazovsky and Shestakov (1994) argue that currently 40 per cent of the population of the former Soviet Union is living in areas characterised by 'acute ecological situations' that are adequately described as desertification. But they also admit that migration from such areas is not new, including as 'desertification-induced migration' such movements as the migration of Mongolian tribes northwards in the second century BC due to drought, and the removal of population from the Khoresm oasis in the first century AD after the invasion of nomadic tribes, which destroyed irrigation systems. This notion of 'environmental refugees' hardly tallies with arguments about recent destruction of the ecological balance by modern society; rather, migration is again perhaps better seen as a customary coping strategy. In this sense, movement of people is a response to spatio-temporal variations in climatic and other conditions, rather than a new phenomenon resulting from a physical limit having been reached.

For the 'environmental refugees' thesis to be plausible in the Sahel and other semi-arid regions, what is required is not simply evidence of migration from what have always been harsh, marginal environments; rather evidence is needed of an increase in migration at times, or in places, of more severe environmental degradation. Such a process is hinted at in discussions of 'stress migration' in the Sahel, one of five phases of response to famine identified by Cutler (1984), the others being sale of stock, wage labour, borrowing of cash or food, and the sale of valuables. Yet as Pottier (1993) observes, there are a number of analytical question marks over developing typologies of responses to famine, and especially over assuming that these occur in a sequence, the last and most severe of which is migration. For Pottier, migration is not an 'end result' which can be labelled simply as a 'problem', but often forms part of the solution to famine for those concerned. In each case, the dynamic causes and consequences of migration need to be investigated, not assumed. Nor are migrants from drought necessarily 'refugees' even in the broad sense of the word. Indeed, Turton and Turton (1984: 179) reported how the Mursi of Ethiopia responded to the 1970s drought through a strategy of 'spontaneous resettlement' in which they systematically avoided distributed relief at institutional feeding points – which might 'have turned large numbers of Mursi into permanent refugees in their own country'.

Some of the evidence that does exist specifically on migration responses to environmental stress points at least in part in the opposite direction. Thus a study by Findley (1994) of emigration from the Senegal River valley in Mali shows that during the drought of the mid-1980s, migration actually *declined* rather than increased. In turn, there was a clear reason for this, since to migrate requires an initial cash investment to pay for travel and associated expenses on arrival, and an economic downturn reduces the ability of families to make such an investment. However, there was an interesting nuance to this finding, in that whereas mainly male *migration* (defined as departure for a period of six or more months) declined, the process of *circulation* (defined as departure for less than six months, and involving many more women and children) did increase during the most severe period of the drought. In a similar vein, Davies (1996) talks of the difference between 'coping' strategies (such as temporary circulation) and 'adaptation' to drought, the latter involving more permanent and irreversible changes in livelihood, and usually an increase in poverty and vulnerability. It is less than clear that migration represents a prominent form of 'adaptation' in the Sahel.

Refugees from rising seas?

Where there is perhaps some more justification of the notion of 'environmental migrants' (if not 'refugees') is in the case of more dramatic and permanent changes to the environment associated with catastrophic events such as floods, volcanoes and earthquakes. Sometimes such natural events involve temporary displacement, as in the case of the Kobe earthquake of 1995, where, according to the *Japan Times*, an initial figure of displaced of over 300 000 fell to below 50 000 within three months of the tragedy. More permanent displacement may be involved though, as appears increasingly likely in the case of the eruption of the Soufriere Hills Volcano in Montserrat in the summer of 1997. This set a first, and largely unexpected, test for the new Department of International Development in the UK, and has resulted in a dwindling of the island's population. What both of these examples have in common was the operation of natural forces largely outside the control or influence of human intervention (even if geographies of the built environment and of social inequality represent prominent factors of risk).

More significant for discussion here, however, is the interaction of such 'natural' and irreversible events with processes of human-induced environmental degradation: in other words, examples where a failure to observe principles of good environmental management and sustainable development can be seen to have contributed to the environmental decline that is at the root of displacement. In this context, perhaps the most significant argument for 'environmental refugees' – and a main plank of the argument of writers such as Norman Myers – is the predicted effect of human-induced climate change, and the impact this may have on sea-level rise, and increased flooding of low-lying coastal areas. A relatively simple assessment is needed to estimate the populations 'at risk', with Jacobson (1988) for example suggesting that a 1 m rise in sea level could produce up to 50 million environmental refugees. Myers again quotes a higher figure, with a forecast of 150 million environmental refugees by 2050 (Myers, 1993b: 191), and it is this figure that is used by the Intergovernmental Panel on Climate Change (IPCC), the UN scientific body responsible for reviewing the causes and impacts of climate change, in its calculation of the costs of not responding (Bruce et al., 1996). Myers (1996) has subsequently put the potential number at 200 million environmental refugees from sea-level rise alone.

None the less, the question of predicting how many people might be forced to leave their homes as a result of shoreline erosion, coastal flooding and agricultural disruption linked to climate change is far from being straightforward. In particular, although Myers identifies a number of parts of the world, including Bangladesh, Egypt, China, Vietnam, Thailand, Burma, Pakistan, Iraq, Mozambique, Nigeria, Gambia, Senegal, Columbia, Venezuela, British Guyana, Brazil and Argentina, as being threatened by 'even a moderate degree of sea-level rise' (Myers, 1993b: 194–5), and is able to point to figures for flood-related deaths in these regions, he does not identify any specific populations that have been forced to *relocate* from flood-prone areas in the recent past as a result of sea-level rises that have already occurred. The point is that there are many potential responses to increased flooding, of which migration is only one. Some of the rural–urban migration that has occurred in areas prone to flooding has been to cities that are hardly better placed to withstand the effects of sea-level rise.

In general, calculating the population 'at risk' from sea-level rise is a long way from predicting mass flight of a 'refugee' nature with its attendant need for international protection and assistance. For example, in a study of response to floods in Bangladesh, Haque and Zaman point out that there are a range of adaptive responses by local populations, which include forecasting, the use of warning systems, flood insurance, relief and rehabilitation efforts. Interestingly, they note that 'in contrast to the English meaning of "flood" as a destructive phenomenon, its usage in Bengali refers to it as both a positive and a negative resource' (Haque and Zaman, 1993: 102). Earlier work by Zaman (1989: 197) stressed how in Bangladesh, 'whilst erosion removes land, new land appears elsewhere', which can be 'used immediately after it re-emerges'. As a consequence, although 61 per cent of his study population in the delta had been displaced, 90 per cent of these households had moved less than 3 km from their original location.

At the same time, scientific evidence for the process of accelerated global warming remains somewhat ambiguous at best (Döös, 1994), characterised by uncertainty and partial understanding. For Roddick (1997: 160), this reflects the fact that 'in competitive academic disciplines operating in a political world dominated by economic interests who want the climate change debate shelved, it is always possible to find scientists willing to question the evidence on which IPCC bases its predictions'. None the less, the doubt surrounding scientific modelling of climate change does call into question, at the very least, the authority of statistics on the likely millions of environmental refugees it will create. What is worse, whatever the basis for forecasting climate change, migration itself has been assumed, rather than there being any real exploration of the social, economic and political forces that will influence its extent and pattern. The somewhat alarmist tone adopted by some environmentalists arguably does little to influence the concrete actions of government, which need detailed predictions of local changes and 'danger thresholds'. And although the 'threat' of migration plays to the weaknesses of governments motivated by anti-immigration rhetoric, it is hardly an acceptable substitute for a detailed assessment of likely outcomes. Where cost calculations have been made of the likely impacts of climate change, these have been heavily dominated by economic reasoning (Shackley, 1997). Thus the IPCC's working group on the economic and social dimensions of climate change adopts the reasoning of Tol (1995), that a completely arbitrary value of three times average annual income per capita can be considered as the cost to regions of departure (Bruce et al., 1996) – despite abundant (if contested) evidence of the role of migration in increasing per capita income in such regions (Stark and Lucas, 1988).

Environmental conflict and refugee movements

In addition to the possibility of a direct link between deteriorating environmental circumstances or dwindling natural resources and induced migration, a further postulated cause of 'environmental refugees', and a link back to the literature on 'political refugees', is the notion that environmental degradation is increasingly at the root of conflicts that feed back into refugee movements. As was noted in Chapter 1, this has become a major theme of the literature on 'conflict studies' as East–West rivalry is no longer a convenient explanation of war, and other factors behind conflict and forced migration need to be found. However, a review of major conflicts that have caused large-scale forced migration during the 1990s, for example, provides little evidence of the generation of environmental 'hotspots' that have developed into war. Thus of the 11 distinct conflicts identifiable from Table 2.1 as being behind 'recent' forced migrations (i.e. since 1990), some, far from reflecting disputes over declining natural resources, could be better described as conflicts in which the protagonists are attempting to control already or potentially rich natural resources. The Gulf War of 1991 occurred as a result of one oil-rich nation seeking to control its oil-rich neighbour; the current war in Sudan is also at least partly about control of oilfields in the south and the building of a canal to open up the southern region (Collins, 1990); whilst Azerbaijan and Kazakhstan, currently undergoing oil-led booms, are hardly the poorest of the former Soviet republics. Both of these latter conflicts, and others, ranging from the former Yugoslavia and the Great Lakes to Bhutan and Burma, might be seen to have more to do with the rise of ethnic (and/or religious) nationalism than overtly environmental conflict.

Of course, in some cases, and particularly in the 'complex political emergencies' of the Great Lakes, Liberia (and Sierra Leone), and Somalia, environmental issues can be seen to have some relevance in the development of hostilities, and a case can be made that environmental degradation forms an important root cause of the conflict. In Rwanda, an extreme position is put by Diessenbacher (1995: 58), who argues that overpopulation not only caused the genocide in Rwanda, but had 'an exponential effect on other influencing variables'. Although his thesis does not rely on environmental degradation *per se*, but rather the failure of the productivity of the environment to keep up with population growth, a clear link with environmental degradation is identified. In general, the image portrayed in much writing on Rwanda since the genocide is of a poverty-stricken country in which the conflict was somehow linked to the inadequacy and deterioration of the resource base, such that the war was partly a struggle over scarce natural resources.

However, an alternative perspective quite reasonably locates the recent conflict in Rwanda in a political struggle for power, in which ethnicity and access to natural resources were both mobilised as issues by powerful élites (Lemarchand, 1994; Prunier, 1995; Reed, 1996). Equally, the history of the region, and especially the history of colonial policy of 'divide and rule' of populations that had previously lived together over centuries (albeit not always in perfect harmony) can also be seen as highly relevant to the genesis of the conflict (Davidson, 1994; Mamdani, 1996). Indeed, the conflict itself appeared to take on a regional character, not limited to the zone of high population density in the Great Lakes itself (Pottier, 1998). Similarly, in the case of the war in Liberia and Sierra Leone, Richards (1996: 115) reviews the evidence for an environment–conflict link, but concludes that 'no direct connection between deforestation and the war is found'; in essence, although Liberia and Sierra Leone have environmental problems, they do not have environmental crises. Instead, Richards

Table 2.1 Major refugee situations, 1996, by host country and region, country of origin and number of refugees

Host country	Region of settlement	Country of origin	Number of refugees
(a) Recent refugee movements, large-scale flows since 1990			
Guinea	Forest Region	Liberia/Sierra Leone	663 830
Iran		Iraq	579 200
FR Yugoslavia	Serbia	Bosnia-Herzegovina/Croatia	547 843
Tanzania	Kigoma/Kagera	Burundi/DRC	440 666
DRC[1]	N Kivu	Rwanda	423 561
DRC	S Kivu	Rwanda/Burundi	_[2]
Germany		Bosnia-Herzegovina	330 000
Côte d'Ivoire	Danane-Tabou	Liberia	327 696
Armenia		Azerbaijan	200 000
Kenya	Daćaab	Somalia	171 347
Croatia		Bosnia-Herzegovina	165 395
Liberia	Lofa	Sierra Leone	120 001
Nepal	Jhapa/Morang	Bhutan	106 801
Thailand	NE Region	Myanmar (Burma)	104 033
Ethiopia	Western	Sudan	75 743
Russian Federation		Kazakhstan	54 819

Table 2.1 (Cont'd)

Host country	Region of settlement	Country of origin	Number of refugees
(b) Protracted refugee situations, existing prior to 1990			
Iran	South Khorosan	Afghanistan	1 414 659
Pakistan	NWFP	Afghanistan	1 200 000
Sudan	Kassala	Ethiopia/Eritrea	379 774
China		Vietnam	288 805
Ethiopia	Haraghe	Somalia	287 761
Uganda	Arua/Moyo/Masindi	Sudan	223 720
Algeria	Tindouf	Western Sahara	165 030
Zambia	NW Province	Angola	109 623
DRC	Shaba	Angola	108 284 —[2]
DRC	Bas Zaire	Angola	—[3]
India	Ladakh	China (Tibet)	98 000
DRC	Haut Zaire	Sudan	96 529
Senegal	Fleuve	Mauritania	64 030
Iraq		Palestinians	62 635
India	Tamil Nadu	Sri Lanka	62 226
India	Tripura	Bangladesh	53 500

Notes:
[1] Democratic Republic of the Congo.
[2] Included in total for north Kivu, as available statistics do not distinguish the two flows.
[3] Included in total for Shaba Province, as available statistics do not distinguish the two flows.
Source: Based on UNHCR (1997).

argues, the causes of the war need to be sought elsewhere. Regardless of the particular root of the war, an analysis such as that of Kaplan (1994: 46), which links together 'disease, overpopulation, unprovoked crime, scarcity of resources, refugee migrations, the increasing erosion of nation-states and international borders, and the empowerment of private armies, security firms, and international drug cartels', provides little causal explanation but much passion in an 'analysis' that is symptomatic of much of the field.

Elsewhere in the world, and in earlier conflicts, once again, the evidence for environmental pressure or degradation (or indeed population pressure itself) actually causing conflict and forced migration itself is limited. Diessenbacher (1995) suggests that of the 181 wars and civil wars world-wide since 1945, 170 have occurred in places suffering from population explosions. But such an association is not a substitute for causal analysis, and in detail, it is a thesis that all too often breaks down. For example, Lazarus (1991) quotes a report by USAID on El Salvador, which argues that conflict between the government and rebels in the late 1980s has resulted in 'fundamental environmental as well as political problems', but remarks that this is hardly evidence that these problems fuelled a war so much rooted in the international politics of the Cold War. In Mozambique, which saw at least three million people displaced abroad and internally, conflict was again more clearly rooted in the Cold War; here, it is interesting that the overwhelming perception of Mozambican refugees on return after the conflict was that they were going back to a country with unlimited resources and few if any environmental problems. However, it is also quite ironic (and telling) that in one of Africa's least populated countries, pressure of population on resources has probably occurred, stimulated not by high population densities *per se*, but by granting of land concessions to private companies (*cf.* McGregor, 1997; see also Chapter 7). In Somalia, the history of Western (and Soviet) intervention is so long that it is practically impossible to disentangle from the troubled history of this war-torn country (De Waal, 1997). The point is that in conflict, as much as in migration, it is difficult or impossible to isolate particular causes outside the broader context within which these processes develop; indeed, conflict and migration themselves are part of a dialectical relationship with this broader 'context', such that a simple causal link from environmental degradation to conflict to migration is hardly likely to be found.

Environment and natural resources in refugee-hosting areas

Turning once again to the question of refugee and forced migration impacts on the 'environment', rather than environmental causes of migration, Chapter 1 outlined the basis of concern amongst international agencies and commentators, without investigating the extent to which positions adopted are supported by evidence. One of the earliest statements by the UNHCR on why environmental issues in refugee situations differ from concerns elsewhere was made in a report of the Programme and Technical Support Section (PTSS) in Autumn 1991. The report, drafted as part of the UNHCR's submission to the Rio 'Earth Summit', cited three specific conditions of refugee situations, namely:

- the disproportionately high population densities often created in areas of refugee settlement;
- the tendency to site refugee camps in environmentally fragile areas; and
- refugees' lack of any incentive to conserve the environment, because of the trauma of war and resettlement, and the fact that 'the land where they are living is not theirs' (UNHCR, 1991: 1).

Since 1991, these 'specific conditions' have been oft repeated, as well as being added to by subsequent UNHCR-written or sponsored reports, to the point that they are beginning to acquire the status of 'received wisdom'. Amongst additional factors mentioned, Thiadens and Mori (1995) have referred to the uncertainty of refugee situations, which impedes proactive planning; whilst Ketel (1994a) referred to concern about the framework of international assistance to refugees in his environmental assessment for the UNHCR of the refugee camps in Tanzania that year. Ketel calls this 'the attitude factor' of international organisations and NGOs whose mission is to be more concerned with the well-being of refugees than longer-term environmental protection. None the less, the temporary nature of the refugees' stay, high population density that exceeds carrying capacity, and siting of refugee camps in environmentally fragile areas, stand out as issues referred to again and again in the literature.

Each of these arguments is certainly plausible, but simply repeating them is again not a substitute for hard evidence that concerns are justified. For example, in Chapter 1, doubts were raised over the notion that refugees automatically lack incentives to conserve natural resources. High population densities may be created as refugees move across a border, but this is not an inevitable consequence of refugee flight, with counter-examples of refugee dispersal being found in West Africa (see Chapter 6) and elsewhere. Even if high population densities are created, this does not necessarily imply environmental degradation; rather this is a matter for empirical investigation. What is perhaps more important than density *per se* is the association of refugees with large, formal camps. Jacobsen (1994: 15) argues that is the 'most important single factor in determining refugees' effects on the local environment', since camps have a unique set of environmental risks, including pesticides and insecticides (used to control disease-carrying vectors), waste disposal and water problems. Within camps, whatever the regional population density, there is also often a failure to provide sufficient relief inputs, such that refugees are forced to use meagre local resources to meet their subsistence needs.

The siting of refugee camps in environmentally fragile or vulnerable areas is another matter of legitimate concern to the UNHCR and other assistance agencies, although once again, the assumption that refugees are always settled on fragile lands deserves further investigation, and exceptions to the 'rule' can be found. Sorenson (1994: 181) argues that 'settlement camps for refugees may be established in marginal lands, particularly in densely populated countries where citizens are already competing for limited fertile land', whilst Lassailly-Jacob (1994a: 215) observes that 'host governments choose remote and isolated areas to prevent refugees from integrating into local communities and competing with nationals for scarce resources and employment opportunities'. None the less, on a purely practical basis, settlement of refugees often occurs very quickly, such that host governments and communities may have little opportunity to prevent refugees from settling in certain areas. Where settlement is 'spontaneous' (a euphemism for occurring on refugees' own initiative), sites are usually found in border areas regardless of their agro-ecological potential, or alternatively in existing towns and cities, which are sometimes preferred by refugees because of the greater access to employment and income-generating opportunities. Where settlement is 'planned' (meaning controlled by governments or the UNHCR), the most immediate practical considerations are the availability of water, an absence of environmental health hazards, and accessibility to aid deliveries, rather than the quality of the land itself (UNHCR, 1982).

In the longer-term, it is again not necessarily the case that refugees are 'marginalised' onto poorer quality land, or indeed isolated from local populations. For example, the history of refugee settlement in Tanzania since independence has been one of the creation of organised settlements in areas which whilst not of the highest quality, were none the less perceived as having productive potential, and certainly became integrated into the commercial economy (Daley, 1991). Indeed, the agricultural settlements that were established in the 1960s complemented the Tanzanian state's policy of establishing *'ujamaa'* villages, as well as providing a labour force to increase national agricultural production. Similarly, in Zambia, Angolan refugees are settled on land of sufficient agro-economic potential that repatriation is being actively promoted by a government that wishes to allow new settlers into the area (Chisholm, 1996). The Senegalese example cited in Chapter 6 of this book is one where refugees have been allowed to settle on some of the most productive and fertile irrigable land in the country.

One test of the extent to which refugees are truly 'marginalised' onto poorer quality or more vulnerable land would be to identify instances in which, having settled in one location, refugees were then subsequently displaced onto more marginal land as a result of competition with host communities or state action. However, although examples of such secondary displacement can be found, they do not on the whole fit the model of 'marginalisation'. Rather, one of the key reasons for secondary displacement, in Africa at least, has tended to be the desire of governments to move refugees away from border areas, where their presence may be politically sensitive, to locations further inside the host country. Whilst this may have the incidental consequence of moving refugees to less productive or more vulnerable land, cases can be found, ranging from southern Sudan in the 1980s (Harrell-Bond, 1986) to eastern Zambia in the 1990s (Black and Mabwe, 1992), where the eventual zone of settlement did afford an environment in which agriculture was both possible and, for a time, successful.

Impacts of forced migration on natural resources: the evidence

Reviews of available literature provide some evidence of medium- to long-term environmental degradation in refugee-affected areas (Black, 1994a; Jacobsen, 1994, 1997), although the reliability of this evidence varies considerably, and does not appear to have improved greatly over the last two to three years. For example, Kibreab (1996: 22) suggests that 'the available literature . . . abounds with a myriad of impressionistic and unsubstantiated assumptions'. Within the literature, detailed studies are often lacking, with conclusions based frequently on visual observation and consensus rather than measurement – although Kibreab's own work on refugees and environmental issues in eastern Sudan represents a rare and honourable exception.

In searching for evidence of the environmental impacts of refugees, it is helpful to start with the question of what is meant by environmental impacts – or indeed what are the 'impacts' that are most expected, or considered most relevant in refugee-affected areas. Ketel (1994a: 10) refers to direct impacts, which he describes as 'notably complicating daily life . . . (but) seemingly soluble in the short term'; and indirect impacts, which may be more diffuse and so less easy to control. Deforestation, destruction of grasslands and animal life, and deterioration of health, water and sanitation conditions are cited as examples of the former, although these have hardly been easy to solve in practice. A range of other 'indirect' pressures are listed, including pressure on women, reductions in animal fodder, increased soil erosion, losses of water (especially groundwater), potential flooding, a reduction in the number and species of wildlife, and a change in the microclimate, although some of these would appear to be quite 'direct' in their impact on livelihoods. Hoerz (1995a) in contrast simply cites a list of impacts, namely deforestation, loss of grazing land, depletion of 'fall-back resources', degradation of agricultural lands, water consumption and pollution, solid waste, and threats to protected areas, with no obvious common thread. Some 'indirect' impacts not mentioned by Ketel or Hoerz might include the loss of tourist revenues or hunting fees to the host community and/or host authorities if animal numbers are reduced, or the loss of royalties on timber, each of which may themselves be relevant to future environmental management strategies.

In order to give some structure to the discussion of impacts, it is important to be more systematic in their identification, and to classify these in a way that is useful for either understanding or policy intervention. Bloesch (1995: 41) attempts the latter by noting that it is possible to distinguish 'immediate' impacts (for example, impacts on resources required for shelter) from 'gradually increasing' impacts, such as those caused by demands for cooking, heating, water or fodder for animals. This is a useful distinction, and reflects the later division of this book into separate chapters on refugee emergencies and more medium-term concerns, although it should be recognised that 'gradually increasing impacts' start at the time when refugees first arrive in a country. None the less, Bloesch's division remains one that distinguishes the type of activities or demands that cause environmental impacts, rather than the nature of the impacts themselves.

In this section, though, a different division is used, based on three key elements of the natural environment that are important for human (or animal) survival, namely impacts on flora and fauna, impacts on soil resources, and impacts on water resources. This follows a similar division made by Jacobsen (1994), who cites deforestation, land degradation and water degradation as three key areas of concern, although she also considers the costs and burdens placed on refugees and hosts, impacts on pastoralists and women, and impacts on food security as relevant factors. Certainly deforestation can be seen as a major concern of all of the authors that have written on this subject, as well as of agencies working in the field (Black, 1994b; Ghimire, 1994), and represents a primary form of change in vegetation or flora. However, the various human impacts mentioned by Jacobsen are excluded from discussion here, not because they are considered unimportant, but because they form a different type of impact: arguably they are a consequence of environmental degradation, rather than a direct *environmental* impact of the presence of refugees.

In the sections that follow, there is a focus on examples of human-induced accelerated environmental change which are considered of potential long-term importance to environmental sustainability at a local level. The emphasis is firmly on land-based 'natural resources', rather than 'environment' in its broader sense, and so a further element – impacts on air quality – is also not considered in detail here. This omission is not because impacts on air quality are thought unimportant, but because it is doubtful that forced migration can have a significant *long-lasting* impact on air quality or the atmosphere. It is worth noting though that air pollution can occur in certain situations in the short term. For example, Lamont-Gregory (1995a: 14) calls for more attention to be paid to 'the association between exposure to raw biomass smoke, acute respiratory illness and the death of malnourished children' in refugee situations.

Changes in flora and fauna

Changes in the vegetative cover of a host area are perhaps the most obvious and visual impact of the presence of refugees, for a number of reasons. On arrival in a new area, refugees create demand for both fuel and construction materials, and in much of the developing world, this is likely to be met from local resources of wood. In addition, there is demand for land for the sites of refugee settlement, and in many cases for agricultural land, and it is likely that fallow or forest land will be cleared for these purposes. Jacobsen also cites the impact on vegetation of livestock brought by refugees, whilst she describes in detail the process of commoditisation of forest resources:

> as demand for firewood increases, markets become established and more distant woodlands are affected; markets for other natural resources like thatching grass or water also emerge; people sell firewood or exchange it for food rations during times of food insecurity, i.e.

during the 'hungry season' after the harvest when food stocks are depleted. (Jacobsen, 1994: 6)

A similar process is described by Hoerz (1995a), who notes the increased cutting of live trees for building material, fencing material, charcoal, and for sale to urban markets. He also notes the 'pull effect' that refugee settlements and camps may exert, citing examples from Kenya and Nepal of camps that had attracted additional local populations after they were established.

Jacobsen echoes the conclusion of other writers when she concludes that deforestation in particular is 'the most serious environmental problem associated with refugees' (Jacobsen, 1994: 5), although she provides little direct evidence to support this conclusion. None the less, evidence of deforestation is cited in a range of studies of refugee-affected areas, in areas ranging from tropical rain forest to dry tropical forest, savannah and steppe. In Somalia, both Young (1985) and Orr (1985) noted visual evidence of the total destruction of remaining areas of isolated woodland over a distance of several kilometres surrounding refugee settlements in a variety of regions. A study by Allan (1987) of the impact of Afghan refugees on vegetation resources in north-west Pakistan suggested on the basis of field observation, analysis of air photographs and satellite images that 'serious' deforestation had occurred in certain areas. Although in some parts of Pakistan Allan observed that virtually no environmental impact had occurred, in such cases there was also very little cultivated land, forest or pasture, and it was seen as necessary for refugee relief agencies to supply tents, food and kerosene from outside the region. More recently, in Bangladesh, a report by the United States Committee for Refugees (USCR, 1995: 20) concluded that the influx of Rohingya refugees from Burma since 1991 had 'indisputably had a negative impact on the environment', with deforestation highly visible around the refugee camps, as a result of the lack of alternative fuel sources.

In Malawi, several studies have pointed to potentially serious environmental impacts linked to large-scale refugee flows. For example, both Wilson et al. (1989) and Tamondong-Helin and Helin (1991) identified loss of woodland as a problem, whilst studies quoted by the World Food Programme (WFP, 1992) and the German agency, Gesellschaft für Technische Zusammenarbeit (GTZ/UNHCR, 1992) also drew attention to rapid deforestation as a result of the burning of wood as fuel by refugees. The latter two studies used a method characteristic of many others in determining impacts on forest resources, by extrapolating figures for average per capita woodfuel consumption by refugees to calculate the volume of wood required, and hence the area cleared in order to meet demand. Using this method, they suggest that between 500 000 and 680 000 m^3 of forest were cleared each year for cooking and heating whilst the refugees were present in Malawi. Similar estimates by the government of Malawi put the figure at 1.3 million m^3, or at a stocking density of 30 m^3 ha^{-1}, some 43 333 ha of forest lost each year (GOM, 1992).

However, it is important to place these and other studies in context. For example, refugee flows are not the only factor causing deforestation:

in the case of Somalia, according to Young (1985), 'a great deal of vegetational degradation had occurred before the refugees had arrived', whilst in Pakistan, Spooner (1987) suggested that some refugee-affected areas had been in their current 'degraded' state for over a century. Allan (1987) has argued that not only had deforestation already occurred in refugee-affected areas, but local entrepreneurs also took advantage of the confusion over land rights caused by the presence of the refugees to engage in illegal logging of timber on a large scale. The same situation is cited for Bangladesh by the USCR (1995: 20), which notes 'a considerable flow of black market hardwood that is not controlled by Rohingya (refugees) but rather by large Bangladeshi-owned lumber yards'. Similar examples can be cited from both Honduras (Girard, 1992) and northern Thailand, where areas affected by refugees have also been subject to extensive commercial logging of timber. Indeed, along the Thai–Burma border, such mechanised logging has formed a part of the government's counter-insurgency strategy, and has itself contributed to population displacement (Ashley, 1992; McGregor, 1993).

At the same time, care must be exercised concerning estimates of deforestation based on per capita consumption. Such an approach has been widely criticised, with Crewe (1997), for example, noting that predictions of complete deforestation in some countries within a decade, made in the 1980s, have proven completely false. Crewe and others have pointed out that such estimates generally do not take into account what a sustainable yield of firewood for an area might be, whilst partial felling, collection of dead wood, and use of wood from agricultural plots are usually preferred by local users, both on grounds of ease of collection, and suitability of the resulting firewood. Given the complex patterns of collection of different kinds of biomass for cooking, the estimates of areas deforested or biomass 'lost' in refugee-affected areas may also be simply incorrect. In the case of Malawi, some of the demand for wood was also met from felling inside Mozambique (Wilson et al., 1989), whilst it was common for wood to be gathered within agricultural plots. Thus after the refugee emergency had ended, a detailed field-based study by the NRI (1994) concluded that in so far as it was possible to distinguish clearance of forest by refugees rather than local Malawians, this was primarily to open up land for settlement and agriculture, rather than in searching for woodfuel.

It is also worth pointing out that an emphasis on 'deforestation' may be somewhat misleading in specifying the true nature of the problem. Even if it is true that in refugee situations, the rate of cutting of trees exceeds the rate at which biomass is replaced by annual growth, this does not necessarily imply total or permanent removal of forest. Much will depend on what happens when, or if, the refugees leave the host area after repatriation. Natural regeneration may quite rapidly replace the tree cover that was initially lost. Moreover, across Africa, there is increasing evidence that areas presumed to be 'primary' or 'undisturbed' forest have in fact been cut in the past. For example, an analysis of documentary sources and oral histories by Fairhead and Leach (1995) suggests that across West Africa, many forested areas have been cultivated in the past, but have subsequently 'recovered' and are today largely indistinguishable from 'primary' forest.[1] If it is accepted that ecosystems are inherently dynamic, disturbance to ecosystems caused by refugees need not be seen as necessarily negative, and may lead to the development of new vegetation types that are more useful, or more productive, than those that were present before. Examples might include the conversion of forest to agriculture or pasture for grazing. Fairhead and Leach (1996) even found in Guinea that human settlement can enhance the quality of subsequent forest regrowth: for example, it was found that the seeds of some tree species need to pass through the gut of a goat in order to germinate.

Forms of vegetation change other than 'deforestation' may also be of interest and/or concern, although these have tended to receive less attention than the loss of trees. From an economic point of view, excessive pressure on grazing land may result if refugees bring with them large numbers of livestock. As Hoerz (1995a) notes, this may have a severe impact on the livelihoods of local pastoralists, whilst contributing to soil erosion through loss of ground cover. However, Hoerz fails to cite a concrete example of this occurring in a section of his report on 'loss of grazing land', whilst Jacobsen (1994: 13) cites 'rangeland degradation in the north' of Somalia and 'overgrazing in the south' without providing the source of this observation. Indeed, the two examples that Hoerz does cite both point in the opposite direction, thus reflecting the point above about conversion of forest land. In Zimbabwe, a post-refugee environmental assessment has described how the grass cover of the refugee-affected area actually improved as a result of a reduction in woody species competition (FCC, 1994). In neighbouring Zambia, both Black and Mabwe (1992) and Lassailly-Jacob (1993) report how bush clearance reduced the incidence of tsetse fly and improved the prospects for cattle grazing. Similarly, Kibreab (1996) reports the reduction of vegetation cover in areas of eastern Sudan settled by refugees as also minimising the problem of sandfly infestation, effectively making the area habitable.

In addition to changes in overall vegetative cover, the presence of refugees may also lead to a loss of particular species of either flora or fauna, with consequences of both local and more global importance. Hoerz (1995a) describes the 'depletion of fall-back resources', in which he is concerned with pressure on specific plant and animal species that are valued as foods by local people. To this can be added species that are of medicinal or ritual value, as well as species or habitats that are threatened globally, and which are considered in need of protection. Research by Wilson et al. (1989) in Malawi, McGregor et al. (1991) in Swaziland, and Black (1994c), Spitteler (1992) and Lassailly-Jacob (1993) in Zambia has pointed to the significance of gathered products (such as plants with edible leaves, tubers, fruits, mushrooms, a variety of insects, fish, rodents and larger game) in refugees' diets and income-earning strategies, and has found evidence of these resources being under pressure. But this does not necessarily mean that species diversity was lost or even threatened in the longer-term – a question that could only be addressed now with follow-up research.

For example, in Zambia, local people and assistance agencies alleged that refugees used bush fires to gather rodents and insects, and cut down whole trees to collect honey and caterpillars (Lassailly-Jacob, 1993), with wider effects on the ecology of the area. Although these allegations were largely unsubstantiated (Black et al., 1990), it does seem that within a few years of the refugees' arrival, many bush products were becoming very hard to find (Sullivan, 1990; Spitteler, 1992). However, since the refugees' departure in 1994, pressure has been significantly reduced on the area of settlement, providing an opportunity for recolonisation by both plant and animal species. Meanwhile, species need not be of economic value to be of conservation importance, with concern over threats to certain species of beetle in the forest reserves in Guinea being a case in point (Bourque and Wilson, 1990). Of course, some endangered species – such as the mountain gorillas of the Virunga National Park in the Democratic Republic of the Congo, and in neighbouring Rwanda and Uganda – do now have considerable economic value as well due to their significance for tourist revenues. Thus far though, it is not suggested that any plant or animal species has become extinct as a result of refugee settlement.

Soil degradation

Whilst much international attention has focused on deforestation and other vegetational changes, both in refugee-affected areas and beyond, the question of soil degradation has received relatively less attention. This partly reflects the fact that less evidence is available to indicate that impacts on the soil itself, rather than on the vegetation that grows on it, is a serious problem in refugee-affected areas, although the potential for such degradation is real. Soil degradation is defined by the UNEP (1992: 11) as 'human-induced phenomena which lower the current and future capacity of the soil to support human life'. This definition ties in with, but also risks being confused with, the broader concept of land degradation, defined by Johnson and Lewis (1995: 2) as 'the substantial decrease in either or both of an area's biological productivity or usefulness due to human interference'.

Various studies of refugee settlement have mentioned soil degradation as a problem in certain areas, although evidence is rarely available in the form of detailed study of environmental processes. In a general discussion of degradation of agricultural lands by refugees, Hoerz (1995a) devotes much attention to the question of the size of agricultural plots allocated to refugees, an issue returned to briefly in Chapters 3 and 6. As befits a discussion pitched at a general level, there is little detailed evidence presented. However, at a regional or country level too, evidence of negative impacts is generally lacking. In Malawi, according to Tamondong-Helin and Helin (1991), 'severe' soil erosion, plus a breakdown of soil ecology and nutrient cycling, is alleged to have occurred. A study for USAID by Long et al. (1990) suggested further that deforestation linked to the presence of refugees was to blame for massive soil erosion after floods in 1989, as heavy rainfall 'washed away soils on deforested land'. But no detailed evidence is cited in either document to support these contentions; and although a visit by this author to Malawi in 1997 provided physical evidence of the destruction of a refugee hospital after a perennial stream dramatically changed its course in flood in the early 1990s (Figure 2.1),

Figure 2.1   Flood damage at Mankhokwe refugee hospital, Malawi, 1996
Source: (photo: Richard Black)

no monitoring of the stream or of deforestation in the surrounding area had been conducted that could have distinguished 'accelerated' soil erosion from natural degradation which might normally be associated with such a semi-arid area. More recently, Biswas and Quiroz (1995) describe serious soil erosion and the formation of gullies in camps near Bukavu in Zaire, citing the lack of terracing and proper drainage channels, and the clearing of vegetation by refugees as the reasons for this erosion, but again, detailed evidence of the extent of the problem is lacking.

One exception to this lack of attention to soil degradation is provided by Kibreab's (1996) study, which focused on the Qala en Nahal refugee settlement in eastern Sudan. Qala en Nahal is described by Kibreab as one of the 'oldest refugee settlements in the world' (Kibreab, 1996: 18), in which there was adequate availability of baseline data to make a scientific assessment of the extent of land and soil degradation – not to mention the author's personal experience of the scheme over a number of years. Kibreab found that there was declining soil fertility on land farmed by refugees, although the irreversibility of this decline was questionable. There was also a decrease in crop yields over a 20-year period, although this was only partly correlated with land degradation, and partly with inadequate and variable rainfall. However, Kibreab is clear that 'overcultivation' was not the product of either ignorance or a lack of commitment to 'sustainability' or long-term environmental conservation on the part of the refugees. Rather, he concludes:

> Most of the differences between (refugees and local farmers) in terms of land-use practices and responses to environmental stresses were attributable to differential access to resources and to other fundamental rights, such as freedom of movement and residence. . . . The problem of over-cultivation, manifested in continuous cropping without fallow periods

and in cultivation of degradation-prone sites, is inextricably linked with the inability of the refugee farmers to expand cultivation beyond the designated areas. (Kibreab, 1996: 314–5)

Indeed, he goes on to note that refugees have adapted to the constraints on agriculture by investing in livestock, which have the benefit of being transportable to Eritrea on their return. In turn, there is little evidence of overgrazing, as opposed to over-cultivation, which Kibreab argues is precisely because no limits are placed on migration of livestock, and so a sensible management system can be put in place by herders.

Detailed analysis of soil fertility on land cultivated by refugee and local farmers has also been carried out by this author in the Republic of Guinea in collaboration with Mohamed Sessay and Faya Jean Milimouno (Black et al., 1996). Once again, to the extent that soil degradation was found to be a problem, this was associated with refugees being forced to cultivate land that had experienced a shorter fallow period, although in this case, the land was allocated by local farmers rather than by the government. Thus levels of organic matter and nitrogen in particular were found to be low on plots that had been cultivated after a three-year, rather than a more normal seven-year fallow, with this apparently having a significant impact on yields. None the less, there was no evidence of declining soil fertility being irreversible, or of physical soil erosion, except in new settlement areas where roads had inappropriately been built perpendicular to steep slopes. Nor were refugees found to be using cultivation practices that were any different to local farmers, with the exception of second-year cassava plots similar to those described by Leach (1992) for Sierra Leone (see Chapter 1).

In addition to these detailed studies, it is also possible to make a very broad-brush assessment of the *risk* of soil degradation in areas affected by refugees, drawing in particular on existing databases such as the Global Resource Information Database (GRID) established by the UNEP in 1985, information from which is synthesised in the *World Atlas of Desertification* (UNEP, 1992). This atlas provides an indication of both climatic and population variables, and the severity of various types of land degradation, as well as the extent of deforestation and other pressures on environmental resources. An initial classification of these areas is made in Table 2.1, which divides the 32 separate flows of over 50 000 refugees identified in Chapter 1 according to whether they are recent (chosen here arbitrarily as occurring in the 1990s) or whether they represent protracted refugee situations, at least in origin. Although recent flows in the Great Lakes and the former Yugoslavia have grabbed world headlines, it is remarkable to note how long-term some continuing refugee situations still are: for example, despite some repatriation, over 2.6 million Afghan refugees are believed to remain in Iran and Pakistan, having been there in many cases since 1979; whilst the Angolan influx into Zambia and the Democratic Republic of the Congo started with the struggle against Portuguese rule after 1961, although there have been flows in both directions across the border since then, as conflict in Angola intensified and receded. Table 2.2 then further classifies these situations on the basis of whether refugees

Table 2.2 Major refugee-affected areas, by nature of settlement and length of time refugees present

Longevity	Camps only	Local settlement[1]	Urban
Recent	DRC (N Kivu)	Côte d'Ivoire	Armenia
	Ethiopia (western)	DRC (S Kivu)	Croatia
	Kenya	Guinea	Germany
	Nepal	Iran (Iraq)	Russian Federation
	Tanzania	Liberia	FR Yugoslavia
		Thailand	
Protracted	Algeria	DRC (Bas Zaire)	China
	Ethiopia (Haraghe)	DRC (Haut Zaire)	Iraq
	India (Tripura)	DRC (Shaba)	
		India (Ladakh)	
		India (Tamil Nadu)	
		Iran (South Khorosan)	
		Pakistan	
		Senegal	
		Sudan	
		Uganda	
		Zambia	

[1] May include some camps.

are primarily in camps (i.e. without access to agricultural land); settled and/or dispersed amongst local populations with such access; or predominantly in urban areas, and therefore unlikely to have a significant impact on soil degradation.

From Table 2.2, it can be seen that the majority of protracted refugee situations now involve settlement in areas where there is some access to agricultural land, although this varies from situations such as Shaba and Bas Zaire, where Angolan refugees are highly integrated into local populations, to areas such as Tamil Nadu in India, north-west Pakistan, and eastern Sudan, where a significant proportion of the refugees either remain in camps, but have access to land outside, or are unassisted refugees, living outside these camps. Only three protracted refugee situations still involve settlement exclusively in camps: in Algeria, the relative permanence of camps for refugees from the Western Sahara relates to the barren nature of the land into which refugees have moved, whilst in India, camps for Bangladeshi refugees from the Chittagong Hill Tracts have been retained in an area with a high local population density. In eastern Ethiopia, the situation is more complicated, with major flows of refugees from Somalia in 1988, and again in 1991, accompanied by returning Ethiopian refugees, and local people displaced by drought. Here again, refugees have moved to an area largely devoid of sufficient natural resources to sustain them, and camps have been established, although a 'cross-mandate' approach to assistance adopted by the UN in the early 1990s is designed to provide assistance for all displaced groups wherever they are, rather than focusing on the maintenance of camps.

In Table 2.3, the severity of soil degradation in general and the severity respectively of erosion by water, wind, chemical deterioration of soils and physical deterioration of soils are shown for each refugee situation where refugees are predominantly located in rural areas. Five locations are identified as having a 'very high' level of soil degradation, although for quite varied reasons: in eastern Ethiopia, this is related to the combined effect of water erosion and chemical deterioration of soils; in eastern Sudan, physical deterioration of soils is the major problem; whilst in south Kivu in the Democratic Republic of the Congo, water erosion is the most significant. More generally, water erosion, which includes sheetwash, rilling, gullying and piping, can be seen as the major process in terms of soil degradation, with highly affected areas being concentrated (though by no means exclusively) in semi-arid or steppe regions. In contrast, wind erosion, including soil deflation, transport and deposition, is identified as a problem only in a small number of situations, whilst physical deterioration of soils, including sealing, crusting and compaction of topsoils, sodification, waterlogging, aridification and subsidence of organic soils, is significant only in southern Iraq, eastern Sudan and along the Senegal River in northern Senegal.

Overall, caution is required in interpreting the results of this analysis, which is based on a fairly crude measurement of a limited set of environmental conditions. For example, individual refugee-affected areas may cover more than one climatic zone as defined by GRID, in which there may be greater or lesser levels of degradation. High levels of local degradation may also not be detectable in analysis on a global scale with a relatively coarse resolution. Severity of soil degradation is assessed by GRID on a combined measure of the degree of degradation, and the percentage of the mapping unit affected, based on an assessment by country-level 'experts' rather than necessarily any accurate maps. Quite apart from the room for error involved in such an exercise, refugee settlements may be concentrated in areas with higher or lower levels of existing degradation than within these arbitrary mapping units as a whole. In addition, in some cases, the classification of the severity of soil degradation here does not correspond with independent reports of the severity of environmental degradation. Thus the high and increased pressure of population on resources whilst refugees were present in Malawi was seen by a number of authors as contributing to environmental degradation, even though the level of soil degradation shown by this study is only 'medium'. Similarly, a report by Gallagher and Forbes Martin (1992) notes that Somali refugees in Kenya 'have settled in some of the most environmentally fragile parts of the country', even though the level of soil degradation shown here for this region is 'low'.

Pressure on water resources

A third area of environmental degradation potentially associated with an influx of refugees to an area concerns pressures placed on the supply and

Table 2.3 Major refugee-affected areas, showing degradation type and severity

Host region	Soil degradation	Water erosion	Wind erosion	Chemical deterioration of soils	Physical deterioration of soils
Ethiopia (Haraghe)	Very high	High	–	High	–
DRC (S Kivu)	Very high	High	–	–	–
Sudan	Very high	Medium	–	–	Very high
Thailand	Very high	Medium	–	–	–
Iran (Khuzestan)	Very high	–	–	Very high	Very high
Senegal	High	High	Medium	Medium	High
Pakistan	High	High	Medium	–	–
India (Tamil Nadu)	High	High	–	–	–
Nepal (SE)	High	High	–	–	–
Iran (South Khorosan)	High	Medium	Medium	Medium	–
Uganda	High	Medium	–	Medium	–
DRC (N Kivu)	High	Medium	–	–	–
Liberia	High	–	–	High	–
Tanzania	Medium	Medium	–	Medium	Low
DRC (Shaba)	Medium	Medium	–	Medium	–
India (Tripura)	Medium	Medium	–	Medium	–
Zambia	Medium	Medium	–	Medium	–
DRC (Haut Zaire)	Low	Medium	–	Medium	–
Guinea	Low	Low	Low	Medium	–
Côte d'Ivoire	Low	Low	–	Medium	–
Kenya	Low	Low	Low	–	–
Ethiopia (Western)	Low	Low	–	–	–
DRC (Bas Zaire)	–	–	–	High	–
Algeria	–	–	–	–	–
India (Ladakh)	–	–	–	–	–

Source: Based on UNEP (1992).

quality of water. Demands on water supply in refugee camps and settlements may create water shortages for both refugee and host populations, which in some extreme situations may necessitate the supply of water by tanker from elsewhere, as was the case in Kibumba camp in Zaire from 1994 to 1996, and has been necessary in the Saharoui refugee camps in Algeria for over 20 years. The lack of time available for planning in emergency refugee situations may also lead to the drilling of wells equipped with oversized pumps, which draw water more quickly than the well can be recharged (UNHCR, 1991). In addition, impacts on water supply and quality may have considerable effects of relevance to the health of refugee and host populations. Most obviously, whereas refugees and locals may be able to respond to depletion of food or energy resources by utilising alternative foods or fuels, no such substitution is feasible for water, without which death occurs quite rapidly. As such, where pressure of population on water supplies becomes acute, the only solution is to find a new source of supply, which may involve further population movement. Large concentrations of people also produce large quantities of human excreta and other waste, which if not properly treated, may lead to pollution of groundwater and soils. This was raised as a significant problem in Goma, Zaïre, in 1994–96 (Biswas and Quiroz, 1995), and is returned to in Chapter 5.

Depletion of water supply, like impacts on air quality, might be seen as a temporary phenomenon, in the sense that stocks may be replenished once demand is reduced after refugees return home. None the less, long-term environmental problems may result in certain circumstances. For example, where the water supply is from a fossil aquifer, as may well be the case in arid and semi-arid regions, replenishment will not occur or will be very slow, leading to a long-term drawdown on supply. Semi-permanent lowering of the water-table may also lead to land subsidence in the long run, a process alleged by the government of Kenya to have occurred as a result of boreholes being drilled too close together in the camps of Garissa District (Jacobsen, 1996). In coastal areas, a lowering of the water-table may lead to saltwater incursion, and salinisation of the groundwater supply, whilst a reduction of streamflow due to excessive drawing of water from rivers may have the same effect on river water used to irrigate agricultural land. Long-term salinisation of soils may result, although the effects of salinisation may reduce with time.

However, more even than in the cases of deforestation and land degradation referred to in previous sections, there is little empirical evidence that links refugee settlement with long-term degradation of water sources, at least in existing environmental assessments of refugee sites. This may partly reflect the tendency of operational agencies to separate water from other environmental concerns. Conversely, one beneficial impact of refugee settlement, or more specifically of refugee assistance programmes, may be the provision of much-needed boreholes which improve the quantity and quality of drinking water. If such programmes are targeted at refugees and host populations, this improvement should benefit the host area as a whole, a matter that is returned to in subsequent chapters. An interesting example of the impact of waste on surrounding communities is also provided by Benyasut (1990), who reports how, far from representing a pollution problem, the waste from the Phanat Nihom Refugee Processing Centre in Thailand was sold by locals for recycling, whilst organic waste was used as fertiliser on surrounding land.

Conclusion: *what cause for environmental concern?*

An important theme of this chapter has been the apparent lack of evidence, either of analytically distinct environmental causes of forced migration, or of significant negative consequences for natural resource depletion resulting from population displacement. However, stating that there is a lack of evidence of simple causality is not the same as saying that there are no linkages between forced migration and environmental change. A first issue involves the need for better specification of what constitutes environmental change, and why it might be theoretically or practically relevant. This is a task that is dealt with in more depth in the following chapter, in the context of discussing tools for environmental analysis. In addition though, there is also a need to reconsider our approach to analysis of what are often complex and multidimensional linkages between refugees and the environment. Rather than seeking evidence to isolate or 'blame' environment as the cause of migration, or migrants as the cause of environmental damage, we need instead to ensure that relationships between people and the environment, and dynamic changes in these relationships, are considered as a *part* of any analysis of the causes or consequences of human movement, within their wider context.

From a more practical point of view, a second major issue that emerges from this chapter concerns the whole question of which resources are important to conserve, or alternatively, which elements of environmental change associated with forced population movements should or could be the focus of public intervention. In the case of environmentally induced migration, such questions are relatively straightforward to answer: serious impacts are presumably those which contribute to undermining livelihoods (or making land uninhabitable) to the extent that migration is (re-)activated, even though this chapter has sought to question the extent to which a simple equation between environmental degradation and forced migration ever operates in practice. In the case of environmental consequences of forced migration, though, the situation is more complicated.

For example, the UNHCR's own environmental guidelines (UNHCR, 1996), which are reviewed in more detail in the following chapter, provide a different breakdown of 'environmental problems associated with refugee assistance' to that used here, and display some different priorities. They deal in turn with six categories of impact: natural resources deterioration; irreversible impacts on natural resources; impacts on health; impacts on social conditions; social impacts on local populations; and economic impacts. The section on natural resources deterioration tallies most closely with the division of impacts used in this chapter, with impacts on forests, soils and water noted. The UNHCR's second category, irreversible impacts, highlights the fact that some impacts may be more extreme, or more serious, in the sense that neither natural processes of regeneration and resilience, nor rehabilitation measures undertaken by governments or agencies, will be able to return these affected environments to the state they were in prior to the refugee influx. This concern with irreversibility is translated by the UNHCR into a particular concern with the issues of conservation of biodiversity and protection of endangered species, although the addition of areas that are an 'important recreation destination' (UNHCR, 1996: 5) into this category does slightly muddy the water with respect to the protection of certain habitats (see Chapter 3).

Moving on to UNHCR's other concerns, issues to do with impacts of refugee settlement on health are not dealt with specifically here. None the less, noting that they exist emphasises the fact that 'environmental concerns' may be felt both in the long term, as is implied in this chapter, and in the short-term, with serious and even life-threatening consequences for refugee and local populations. Impacts on social and economic conditions, whether for refugees or for host populations, are also not dealt with specifically in this book. To include impacts on the economy and society that are not related to natural resource use (such as the quality of social infrastructure such as schools, hospitals and roads) is seen as stretching the definition of 'environment' across two analytically distinct issues. None the less, it is clearly impossible to discuss concerns relating to natural resource use without at the same time considering social and economic concerns and impacts. Indeed, except in certain extreme circumstances, it could be argued that the main purpose of environmental protection is to ensure the social, economic and cultural survival of human populations (who use natural resources), whilst the conservation or preservation of natural resources inherently has a socio-economic (or indeed political or cultural) dimension.

Finally, these distinctions also beg the broader question of why we are concerned with environmental degradation at all. The case for environmental concern was made in Chapter 1, although some of the competing interests and incentives that have propelled environmental awareness to the forefront in refugee situations were also commented on there. But moving on from this general concern, it is important to recognise that there may be conflicts between local needs for resources, and international concerns over what are seen by some as 'common' resources of global importance. These conflicts are reflected in the discussion in the following chapter, in which practical methods for identifying environmental impacts are reviewed.

1. One objection to this point might be that although tree cover is 'restored', there is a loss of diversity of species characteristic of 'primary' forest. However, this depends on both the type of forest involved and the nature of clearance. For example, in the case of miombo woodland, clearance that does not involve removal of tree roots can enhance rather than reduce species diversity (Tuite and Gardiner, 1990).

CHAPTER 3

Towards an environmental toolkit

The lack of evidence of significant environmental change in areas affected by large-scale refugee movements, highlighted in the second half of the previous chapter, could be interpreted in several ways. It could be argued that this indicates that environmental issues are simply not a problem in refugee-affected areas, although this would seem to fly in the face of the experience of professionals and politicians concerned with the problem. Alternatively, it could be argued that environmental issues and problems are hard to identify, because they are intricately tied up with a range of other complex social and political processes that are occurring across host and source areas. This is a theme that makes more intuitive sense, and is followed up in more detail in case studies in Chapters 5–7. However, a third, and obvious reason for a lack of evidence concerns a lack of attention to data gathering, and it is to this question that this chapter is devoted. First, the extent to which operational agencies systematically analyse environmental issues is considered, highlighting several initiatives to better identify environmental degradation. The chapter then focuses on three main strategies: the use of 'geographical information systems (GIS) and remote sensing technology; the growth of formal environmental assessments at early stages of a refugee emergency; and the development of guidelines for environmental management. Limitations with each of these approaches are identified, and a case made for a more detailed and location-specific approach to environmental problems. Such an approach, it is argued, needs to be both gender-sensitive and based on a participatory analysis in which there is an understanding, even if not a direct application, of longer-term sustainable development concerns.

Identifying environmental impacts: the current situation

Before considering in detail the various strategies that are available to identify environmental impact in refugee situations, or to implement as part of an environmental management plan, it is useful first to consider the (brief) history of refugee assistance agencies' attempts to analyse and deal with the environment. A starting point is provided by a postal survey of 50 operational and funding agencies working in the field of implementing assistance programmes for refugees and displaced persons, which was conducted by the present author in 1993 (Black, 1994b). The survey involved mainly UK-based NGOs, but included some significant non-UK NGOs and intergovernmental agencies represented in the UK. A total of 39 responses were received from this survey, from which 23 agencies were selected for analysis on the basis that they reported involvement in agriculture, forestry, income generation and/or specialist environmental programmes in refugee-affected areas. The survey excluded the UNHCR, which is generally responsible for co-ordinating assistance to refugees, but usually does not 'implement' projects in the field.

The results were interesting, in that they indicated considerable awareness of the importance of environmental issues, but also a lack of practical knowledge of the real extent of environmental problems, or of strategies to deal with it. First, none of the agencies contacted reported involvement in overall planning of a refugee settlement – although this is a job that would normally be the primary concern of the UNHCR anyway. Most reported that prior to implementing an agricultural or income-generation project, they would normally conduct a needs assessment survey of the refugee population; ten said that they would also carry out an assessment of the availability of natural resources, whilst seven said that they would normally carry out (or fund) a soil survey or more formal environmental impact assessment. In turn, 14 of the 23 agencies were also involved in funding or operating programmes designed to reduce environmental degradation in refugee-affected areas, either directly, or as part of a wider programme with other socio-economic goals. The most common form of environmental programme cited was in the forestry sector, including the raising and distribution of seedlings, tree planting, and the development of community woodlots and agroforestry schemes, whilst some agencies were also involved in soil and water conservation and other environmental programmes. A concentration of projects in Sudan and the Horn of Africa was found amongst the UK-based NGOs, although projects were also identified in Afghanistan, Pakistan, Nepal and the Philippines. In many cases, such programmes were relatively small in scope, although this sometimes reflected the small-scale nature of the agency's operation overall.

However, despite involvement in such programmes, only 11 agencies reported that they also had a formal environmental policy or set of guidelines which was applicable to refugee assistance programmes, and in no case was it clear that such guidelines were directed towards specific problem analysis in refugee-affected areas, where situations might differ from more general development programmes. Where guidelines did exist, they varied enormously in their level of detail. For example, one agency commented: 'we simply (not scientifically) ask ourselves what is the impact on the environment'; whilst another stated that 'sustainable development' represented a vital element of programme design overall, but failed to say how this was defined or measured. Only three agencies provided or referred to a specific document or checklist which dealt in detail with environmental issues to be considered in programme planning, whilst one other said that guidelines were under preparation. One agency said that although it had no guidelines of its own, it would expect to follow host government guidelines on the environment.

Since 1993, though, there has been a spate of agency activity in the environmental field of relevance to refugee situations. For example, the Lutheran World Federation (LWF) has developed environmental guidelines for use in emergencies (LWF, 1997); whilst the US-based Care International has financed environmental assessments notably in the Great Lakes region of Africa (ERM, 1994, 1995), which are designed as a blueprint for environmental analysis in such situations. At a global level, perhaps the most important development was the establishment of the post of Senior Co-ordinator on Environmental Affairs in the UNHCR in 1993, and the initiation of a work programme that has encompassed building up a remote sensing and GIS capability, the carrying out of environmental assessments, and the development and implementation of environmental management plans, on the basis of a comprehensive set of environmental guidelines (UNHCR, 1996). These three key areas are considered in detail in the following sections, and the capability of agencies to identify, monitor and respond to environmental problems is assessed. The question of specific environmental interventions by assistance agencies in refugee emergencies is then developed further in Chapter 4.

Remote sensing and GIS

Interest in remote sensing as a planning tool for refugee situations has grown in recent years, with hope expressed that it represents a cost-effective instrument for use in site selection, settlement planning, and environmental monitoring and assessment activities. Remote sensing encompasses the use of both aerial photographs and images from satellite-borne sensors, where available technology has grown considerably in the last two decades. However, whilst aerial photographs and satellite imagery, as well as the associated technology of GIS, are powerful tools that do have relevant applications in refugee-affected areas, it is important not to overstate the capabilities either of current technology or of technologies that might (or might not) be developed in the future. Cost, availability of imagery, and technical criteria place important limitations on what can realistically be achieved, at least at present. Questions might also be raised over whether this 'high-tech' route is the most appropriate one to take.

Remote sensing can be used to infer both the physical characteristics of the Earth's surface and certain aspects of human occupancy of the land. For example, the production of land cover, vegetation and other thematic maps can be greatly assisted by the use of such imagery, and although the need for 'ground truth' (verification by ground survey of the categorisation used in image interpretation) can never be removed, considerable time that would otherwise be spent on expensive land surveying can be saved. The compilation of a new map from remotely sensed imagery is certainly not instantaneous, and the coverage of areas of the world in which a sufficient amount of work has been done already to produce a detailed map quickly is limited. None the less, remote sensing provides the possibility of acquiring relatively quick and cheap thematic maps for use in planning refugee settlements. These features are especially important in refugee sites where there is a need for a speedy response.

Several specific applications of remote sensing stand out as being particularly useful in refugee situations. The first involves the use of a single image to produce thematic maps, for example to identify existing land cover, locate rivers, roads or settlements, or identify salient geomorphological features in an area prior to establishing a new refugee settlement. The use of aerial photographs for this kind of work has a relatively long history, and can provide very detailed maps for use by planners. More recently, satellite images have been used to produce land cover maps in various parts of the world, providing reasonably accurate estimates of the size and distribution, particularly of larger-scale vegetation formations. Thus in many cases, it may not be necessary for refugee assistance agencies to produce their own maps, with national land cover maps already available or being planned by countries such as Tanzania, Uganda, the Philippines and Afghanistan, all host to significant numbers of refugees, returnees or displaced persons.

Beyond the production of maps, other tasks can be completed using single images, though as with maps, this requires trained staff, and 'ground truthing'. For example, by calculating the normalised difference vegetation index (NDVI) from satellite images, it is possible to estimate the woody biomass in an area, which may be useful in assessing supplies of woodfuel or timber. Interest has also been expressed in the option of using remotely sensed images to estimate the size of refugee populations. For example, in a standard text on applied remote sensing, Lo (1986) identifies two methods for estimating populations from such images. One involves calculating the land area covered by a settlement, and multiplying by the population density of a 'typical' settlement as measured by sample survey. In an African context, Lo found this to be accurate for relatively homogeneous settlements and especially those of under 10 000 population, but subject to inaccuracy for larger towns. An alternative involves counting the number of 'dwelling units', and multiplying this by the number in each household, again as measured by sample survey. Here, inaccuracies stem from the fact that not all dwellings contain one household, although the structure of assistance programmes, and aid distribution to 'households', may make this less of a problem in refugee situations.

However, even in the use of a single image to construct a map, or carry out a simple planning or monitoring task, some problems immediately emerge with the use of remotely sensed imagery. Putting aside the point that all use of remote sensing, like the example of population estimation above, involves image interpretation and therefore possible sources of error between that interpretation and reality, perhaps the most obvious

Table 3.1 Commonly used remote-sensing images and their resolutions

Image	Spatial resolution	Temporal resolution	Spectral resolution
AVHRR-NOAA	1 km	Daily	4 wavebands
Landsat-TM	30 m	2 images/month	7 wavebands
SPOT multispectral	20 m	1 image/month	3 wavebands
SPOT panchromatic	10 m	1 image/month	1 waveband
Radar images	Variable	Variable	Variable
Air photographs	1:15 000–1:50 000	Irregular	Variable

problem relates to scale. What is important is the size of a 'pixel', or the area from which a single spectral response is obtained. Put simply, the larger the pixel, the lower the spatial resolution of the image, and hence the less detail is obtained. Conversely, working at higher spatial resolution involves a more expensive and time-consuming process of image analysis per unit area, and in general fewer images are available. For example, widely available satellite imagery ranges from AVHRR-NOAA, which provides daily images of the whole globe but at best at a spatial resolution of only 1.1 km^2, through to Landsat and SPOT imagery, which is at a higher spatial, but a lower temporal, resolution; in other words, there are fewer images per month from which to select (Table 3.1). Much-vaunted satellite images of high spatial resolution on which even individuals can be identified are mostly produced for military purposes and are usually classified.

For compilation of a one-off map, the fact that only one image is available each month is unlikely to be a problem, although the requirement for images without significant cloud cover does reduce availability quite considerably, especially in tropical regions. However, even with a spatial resolution of 20 m, the 'minimum mapping unit' or smallest area that will be identified as a separate feature for multispectral SPOT imagery is still of the order of 1 km^2, too large certainly to identify individual dwelling units. Meanwhile, the availability of a good base map remains important, and the fact that topographic maps are often only available at a scale of 1:250 000 means that this can become the scale of mapping by default anyway (Haack and English, 1996).[1] Clearly the use of the older technology of aerial photographic interpretation can lead to much more detailed maps, but here the choice is likely to be between the expensive task of hiring a plane to take new photographs or using existing images, which are often out of date and may be of highly variable quality or have gaps in coverage.

A review of the utility of remotely sensed imagery for the compilation of resource inventories by Wilkie (1994) makes quite depressing reading, as it is noted that AVHRR images fail to detect small-scale agricultural clearings, which are the primary cause of land cover change, whilst Landsat is too coarse to map land cover in low population density areas. High-resolution imagery without detailed collateral field data on population density and land use provides highly inaccurate estimates. None the less, imagery from satellite remote sensors has been used by the UNHCR in some refugee situations, with notable examples being the former Zaïre and Tanzania. Though in its infancy in terms of implementation in refugee situations, the potential of satellite images, which in some cases may be the only up-to-date 'documents' of the landscape available, is clearly worth exploring. Meanwhile, advances in technology, such as the increasing availability of synthetic aperture radar (SAR) systems, which operate in microwave wavelengths and can thus penetrate cloud cover, promise further improvements in the future (Curran and Foody, 1994; Parry, 1996).[2]

Opportunities also exist through the integration of remotely sensed data in a GIS to produce a database of environmental variables that are of use in planning in the longer term. GISs are spatially referenced computer-based systems for the acquisition, storage, analysis and display of data. Some data may be spatially continuous (e.g. altitude), whilst others are held in an 'attribute' database linked to points or areas on a map. Some governments and a number of international agencies have already produced, or are developing, such databases of environmental variables. For example, the Global Resource Information Database (GRID) mentioned in Chapter 2, established by the UNEP in 1985, contains environmental data from the Global Environmental Monitoring System (GEMS) and other databases and is available for use, although much of the information is at a global or continental scale, and is of insufficient detail for planning in a particular refugee situation. Similarly, the World Conservation Monitoring Centre set up by the UNEP, the World Wide Fund for Nature (WWF) and the International Union for the Conservation of Nature (IUCN) holds digital data on tropical forests, wetlands, world heritage sites and protected areas, whilst the FAO has produced a global database of monthly temperatures and precipitation for a 30-year period, which is available on CD-ROM. Two further ongoing projects noted by Curran and Foody (1994) are the Tropical Forest Monitoring Programme of the FAO and the Tropical Ecosystem Environment Observations by Satellite (TREES) programme of the European Union and the European Space Agency, both based on AVHRR images, although neither yet provides accurate and complete coverage that could be used by the UNHCR on an operational basis.

Partly to address this lack of a specialist source of environmental information, the UNHCR has itself been involved in developing a GIS database, which includes the location of refugee camps, physical infrastructure such as roads and settlements, hydrological systems and the location of protected areas. The database, funded by the Japanese government, is designed to provide relevant environmental and other information to planners of refugee settlements. It has a working scale of 1:1 000 000, using mainly FAO and WCMC data for environmental features. Numerous maps have been produced from this work, in which overlays of different kinds of data can be reproduced on a single map, although the majority are at a national scale, limiting their usefulness for local-level planning. They have, however, been used to identify refugee sites which are close to or inside protected areas, tropical forests, world heritage sites or wetlands of conservation importance, and in the Democratic Repubic of the Congo (DRC) and Tanzania, there has also been an attempt to develop a more detailed database, based on SPOT images at scales of 1:50 000 and 1:100 000 (Bouchardy, 1995).

Other maps of particular settlements have also been produced semi-independently of the UNHCR. An example is the work of Cambrézy (1995), co-funded by the UNHCR and the French agency Orstom, on the Kakuma and Dadaab refugee settlements in Kenya, where maps of the settlement have been created using aerial photographs and digitised settlement plans. There is also growing interest in the use of 'global positioning systems' (GPS). These are hand-held devices that can determine the geographical location of the user to between 20 and 100 m. Thus Collins (1996) has used GPSs to map the relationship between cholera, dysentery and a range of environmental variables in settlements affected by internal displacement in Mozambique, whilst Peluso (1995) reports on their use by forest dwellers in east Kalimantan, Indonesia, to map the location of tree products in order to protect them from encroachment by timber extractors and private developers. None the less, the quality of such maps is still only as good as the data that are put in, which must inevitably continue to rely on field data collection of variable quality.

So far, discussion has focused on the generation of maps and databases using remote sensing and GIS which are of use to planners – including environmental planners – because they can provide a reasonably accurate and up-to-date picture of the state of the environment at a particular time. To measure the impact of refugee settlements on the environment, however, it is necessary to have more than one remotely sensed image, so that either a 'before' and 'after' picture of environmental variables can be established, or better still, ongoing monitoring of these variables can be attempted over time. For example, given suitable time-series imagery, it is possible to calculate areas that have been denuded of vegetation over a particular time period around existing refugee settlements, although from such sources alone, it would not be possible to demonstrate a causal link between the presence of refugees and vegetational change.

This is a valuable endeavour, but it is perhaps unsurprising that few time-series studies of this kind have been attempted at present. First, such analysis is likely to be expensive, in terms of the cost of recent imagery and the time and skill needed for interpretation and analysis, placing it beyond the reach of many operational agencies assisting refugees. Problems relating to the availability of imagery, and the computer power needed to store and analyse multiple images, simply magnify those involved in working with single images, although these problems are reducing rapidly with improved computer technology. Meanwhile, because vegetation, for example, changes on an annual cycle, it is important that images chosen for comparison should be taken at the same time of year. Preferably, this should also be a time when it is easier to distinguish between different types of vegetation cover. However, this 'temporal window' when land cover is highly differentiated (an example being the period immediately after land clearance and/or ploughing, when fields are clearly visible) may also coincide with the period of highest cloud cover. Thus land clearance in the tropics often occurs at the onset of the first rains, when it may indeed be difficult to obtain satellite images that are cloud-free.

One early example of vegetation analysis using both satellite and air photographic imagery is provided by the work of Allan (1987) in areas of northern Pakistan affected by refugees from Afghanistan, which showed significant degradation of forests in certain areas. The need for field verification was highlighted in this case by the fact that trees had been lopped from the bottom upwards, leaving the crowns of trees, which appeared intact on remotely sensed images. However, field verification is itself a costly process where large areas are under study, and cannot easily be done retrospectively for images acquired for an earlier time period. Analysis of vegetation change using remote sensing has also been attempted by the author in association with colleagues in West Africa (Black and Sessay, 1997a; see also Chapter 6). An example of the difficulties encountered is provided by the fact that in the Forest Region of Guinea, host to over 500 000 Liberian refugees since 1989, there was not a single completely cloud-free image available on SPOT or Landsat for the period 1988–95, whilst even those images that were relatively cloud-free were concentrated in January, when very little cultivation activity takes place, and bush regrowth makes fields, for example, very difficult to identify. Meanwhile, aerial photographs were of generally poor quality, with the one good series available having large gaps in coverage close to the international border (where the refugees were concentrated) because of refusal of the company that made them to fly over Liberian airspace!

In addition, questions might be raised over the appropriateness of large-scale investment in remote sensing and GIS technology, not only on grounds of cost, but because of its political overtones. Monitoring of the environment from space might appear a politically neutral task in itself, but it fits well with a security agenda that may be more concerned with troop movements than refugee movements. In essence, what is being invested in is a sophisticated surveillance capacity. In turn, use of this capacity might be either positive or negative for both the environment and people, depending on who controls and has access to it in the longer term.

Environmental impact assessments

Whilst remote sensing and GIS can be viewed as potentially useful though also potentially dangerous in planning or quantifying environmental or other aspects of refugee programmes, this section turns to a methodology that is primarily concerned with predicting impacts. *Environmental impact assessment* (EIA) is now a well-established instrument designed to bring environmental concerns into the planning and decision-making process. Its implementation has mushroomed over the past two decades, particularly in the developed world, where a number of countries have introduced legislation requiring EIAs for 'development' projects over a certain size. However, EIA has not been limited to the advanced industrial nations, with the World Bank committed to some form of the procedure for large-scale bank-funded projects (Goodland et al., 1995), and governments such as those in China, India, Indonesia, Malaysia, the Philippines, the Republic of Korea, Sri Lanka and Thailand all now requiring EIAs for certain types of infrastructural developments (Werner, 1992).

The use of EIAs in refugee situations is still very much in its infancy but has been attempted, notably in the aftermath of the Rwandan crisis in 1994, when environmental assessment studies were conducted by several agencies, including the UNHCR, in Tanzania, Zaire, and Rwanda itself. The aim here is not to consider these specific assessments, which are dealt with in more detail in Chapter 5. Rather, in this section, the principles of EIA are considered, and its more general relevance to refugee emergencies and planning of refugee settlement and assistance programmes. In principle, EIA represents a relatively standardised format for predicting and then monitoring the environmental impacts of projects ranging from dam construction to industrial development. Several specific methodologies are in use, each with its own merits and drawbacks in terms of ease of implementation and its value in the planning process. However, a number of common threads can be identified which apply to most or all EIA procedures, which are worth examining in a general sense for their relevance and applicability in refugee situations. For example, whatever the specific methodology used, according to Wathern (1988: 6), an EIA can be defined as:

> a process for identifying the likely consequences for the biogeophysical environment and for man's [sic] health and welfare of implementing particular activities and for conveying this information, at a stage when it can materially affect their decision, to those responsible for sanctioning the proposals.

As such, an EIA can be seen as a tool for use in the decision-making process on projects, and in its original formulation, specifically to be used at the planning and authorisation stage. The outcome of an EIA is normally an *environmental impact statement* (EIS) or 'environmental appraisal', a formal document which states the purpose of the project; discusses alternative options including the planned project, alternative strategies, and the 'no-action' option; provides a baseline description of the affected environment; and then considers in turn the potential environmental consequences. Recommendations can then be made as to which, if any, of the project's alternative plans should be implemented, as well as drawing up an environmental action plan.

Wathern's definition of an EIA has been extended in practice, especially to cover socio-economic effects of projects, and to influence points of the decision-making process other than authorisation, where EIA may otherwise be seen as an 'add-on' process divorced from other elements of project planning. For example, Werner (1992) suggests that EIA can be used both as a decision-making instrument to decide on acceptability of a project and as a planning tool to minimise adverse impacts, assuming that the project will be approved. Clearly the latter function is more appropriate if EIA is to be extended to refugee situations, since the 'no action' option is likely to be unacceptable in terms of the consequent human suffering of the refugee community. Equivalent guidelines for environmental assessment of resettlement projects published by the FAO, although they make reference to the possibility that resettlement may be found not to be feasible, are essentially focused on the choice of different strategies, assuming that a project will go ahead (Burbridge et al., 1988).

Even if EIA is seen as a part of the planning rather than the decision-making process, some problems remain in producing and implementing what would be regarded as a 'standard' environmental assessment. A first problem concerns the timing of the EIA. Given that an EIA would normally be carried out as part of the feasibility study for a project, environmental mitigation measures would normally be expected to be built into the design and engineering phase. However, this would appear to be difficult if not impossible in the case of refugee flows, which are unplanned and uncertain almost by definition. And although use of EIA as a planning rather than a decision-making tool might facilitate its implementation at other stages of a refugee emergency, it is worth noting the comment of Wathern (1988: 6) that 'the greatest contribution of EIA to environmental management may well be in reducing adverse impacts *before* proposals come through to the authorisation stage' (my italics). The problem of timing is compounded if the 'standard' EIA procedures of 'screening' a project, to see if an EIA is necessary, and 'scoping', to identify the most salient environmental problems for further investigation, are to be conducted prior to implementing a full EIA.

A second major problem with the application of EIA methodology to refugee situations concerns the definition of what, exactly, is being assessed. In the case of the construction of a large dam, or a petrochemical plant, it is clear that the EIA is concerned with a *project*, which is likely to have a specified location and relatively easily identifiable beneficiaries and victims. For refugees, however, the situation is different. For example, should an EIA be assessing the impact of the presence of the refugees themselves, or specific project activities designed to assist them, such as the construction of a camp or settlement? In addition, given the almost inevitable timing of an EIA after the fact of refugees' arrival in an area, there is a strong risk that insufficient attention will be paid to alternative policy scenarios, in terms of the implementation of assistance programmes. Of course, these are issues that can be addressed at the 'scoping' stage, but there may be strong institutional pressure against calling into question a programme of assistance that has already been initiated.

These problems are compounded if the EIA is to contain an element of cost–benefit analysis (CBA), as has been advocated, for example, by Hoerz (1995a). Cost–benefit analysis, of which there are a number of variants (*cf.* Markandya and Richardson, 1992), represents a mechanism for quantification of impacts in economic terms, including a monetary valuation of effects even in the absence of a market to determine such values. A similar process can also be used to assess the 'value' of different policies designed to mitigate impacts. Environmental impacts would normally be classified in this case into two types, based on the principles of 'additionality' (i.e. net impacts) and 'displacement' of impacts from or to elsewhere. However, this raises the question again of where to draw the boundaries of an EIA. Since refugee situations involve, by definition, displacement of populations, it might be expected that the majority of impacts (e.g. on woodfuel use or agricultural land) also represent displacements of activity from elsewhere. In such circumstances, the question arises as to whether an EIA should incorporate impacts on the home region or country of the refugees into the appraisal. In reality, this is unlikely to be politically feasible, due to almost inevitable tensions between sending and receiving countries.

Cost–benefit analysis of environmental impacts itself contains many difficulties, notably concerning issues of ecosystem complexity, the comparative valuation of reversible and irreversible impacts, as well as how to value non-economic components such as wildlife, landscape, etc. (Hanley and Spash, 1993). A common method in CBA is 'contingent valuation', in which assessments of the value of a resource are made on the basis of 'willingness to pay', or 'willingness to accept compensation' for its use. However, Jacobs (1991) points out that this works best where people understand the principle of valuation of a resource, and are able to make reference to a parallel market situation; in other words, where the valuation is perceived by the valuer to be 'realistic'. This situation may not be the case in much of rural Africa, for example, and also begs the question of who owns, or has the right to value, a resource. Commercial valuation of an area of land or of a forest resource on the basis of its productive capacity is quite a different matter to valuation of a species on the verge of extinction, in which a 'notional' value might be added to reflect the uniqueness of the resource (Spellerberg, 1992); whilst as Jacobs (1991) notes, any valuation will depend in part on what else the valuer might spend the money on. In a cross-national comparison of attempts at resource valuation, Abelson (1996) notes some success in putting a value on deforestation in Kenya, although this involved several months of fieldwork in determining market/proxy prices for numerous forest products; whilst in Bolivia, valuation of deterioration in soil quality was only possible because of the 15-year presence of a British Tropical Agricultural Mission. He goes on to note that 'not surprisingly, it proved impossible to value a large number of other ecological impacts' (Abelson, 1996: 271), whilst where valuation was attempted, it was often impossible to capture the multidimensional nature of natural resource values. Perhaps the most difficult problem with regard to contingent valuation raised by Abelson though was that it is difficult or impossible to assign values 'where there is substantial (and often bitter) dispute about income distribution or property rights' (Abelson, 1996: 275), something that is all too often the case in refugee-affected areas.

Beyond the valuation of resources, and the application of cost–benefit analysis, to arrive at a statement about the overall environmental effect of a project, four main EIA methodologies are currently in use which come to more open-ended conclusions about likely environmental impacts. The simplest approach is to develop an 'environmental checklist' of possible impacts, and assess whether each is likely to be significant. This acts as a pointer to areas in which environmental policy should be formulated. Although easy to understand, this approach suffers from the need for a checklist to be exhaustive, and therefore somewhat unwieldy, if significant impacts are not to be overlooked. A second, and more refined approach, involves the generation of a 'matrix' of project activities and their respective impacts (the most well-known being the 'Leopold matrix'; see Leopold et al., 1971). This helps to identify project activities, as well as environmental phenomena, which need to be the target of policy, although the result is often even more unwieldy. The approach has been used by Ofori-Cudjoe (1990) to analyse forced resettlement as a result of construction of the Kpong Hydro-Electric Dam on the Volta River in Ghana, although this was an *ex post facto* evaluation that ultimately had no bearing on policy. A third approach to EIA is the network approach, in which a network of interrelationships is constructed for the ecosystem under study. This can be particularly useful in identifying indirect impacts, which may be overlooked in simple checklists. In turn, a fourth approach involves the drawing up of overlay maps of the spatial distribution and intensity of particular impacts, in order to identify particular zones in which there is intense environmental pressure.

Each of these methods has its advantages and disadvantages, whilst many more methods are available: Caldwell et al. (1982), for example, identified 15 different categories of EIA which were used in a survey of four federal agencies in the USA, whilst it was common procedure to tailor EIA procedures to the particular type of environmental problem under investigation. However, what is arguably common to all is the absence of specific guidelines on how each individual 'impact' is to be assessed, if it is not to be by economic valuation. In this sense, EIA is not so much a methodology for calculating or predicting environmental impact, but more a framework within which such calculations can be integrated and presented. And however sophisticated the model for this presentation, it is clear that the 'accuracy' of the EIA is only as strong as its weakest link.

This would not be such a problem if it were not for the fact that a number of 'weak links' in calculating environmental impacts in refugee-affected areas can be identified. In practice, for example, most if not all of the impacts identified in an EIA are likely to be based on estimates of resource use rather than actual measurements, because the time required for accurate measurement is too long. These estimates then need to be projected into the future, based on a series of assumptions that may or may not be valid in the dynamic circumstances of a refugee emergency. The case of estimating woodfuel consumption based on extrapolations of data on per capita use cited in the previous chapter is a case in point. Another example is provided by soil erosion, which as Millington (1992) points out, is a function of rainfall erosivity, run-off erosivity, soil erodibility, slope parameters, vegetation cover and management and cultivation practices. It is unlikely that all – or indeed any – of these parameters would actually be measured in the field, with estimates being made instead based on visual observation of what appears to be current erosion (which may in fact be relict features), or severity of risk inferred from existing information. Indeed Thanh and Tam (1992: 6) remark from personal experience of conducting EIAs, 'with regard to data and information, we tend to rely on what was said before, without assessing its reliability or validity.'

Such omissions are less serious if a further aspect of standard EIA – monitoring of the actual effects of the project – is included as part of the EIA process. However, even in the more stable policy environment of development projects, Kakonge (1994: 297) concludes that 'Africa has minimal experience with EIA monitoring', noting that monitoring activities are rarely followed up because their impact is seen as slight, and (once again), 'it is difficult to acquire adequate data' (Kakonge, 1994: 299). At present, the history of use of EIAs in refugee situations is too short to come to a conclusion about whether effective monitoring activities have been carried out. The process of implementing the conclusions of an EIA also requires political will, and preferably a national environmental policy framework, in order to be effective. These conditions are all too often absent in developing countries, often due to scarcity of resources and skills (Kennedy, 1988), although important developments are now taking place in a number of countries in terms of the development of national environmental action plans and the strengthening of environment ministries that could take on a monitoring and co-ordination role.

Despite potential problems with monitoring and implementation, it is worth noting that in the process of writing an EIA, far from there being a vacuum of information, there may be plenty of (or even too much) information. For example, any EIA will of necessity be drawn up in the context of existing reports and documents, and if the EIA team has sufficient time and/or knowledge of an area, this may include any significant existing literature on environmental or socio-geographical problems and issues. One particular source of information identified by Guha-Sapir and Salih (1995) is that provided by early warning systems (EWS), which have been developed since the mid-1980s to provide early warning of droughts and famines. These are now relatively common in the Sahel region of Africa, but have also been developed elsewhere. Guha-Sapir and Salih note three aspects of EWS which are of relevance to EIA, namely risk assessment, and the detection, evaluation and prediction of environmental hazards; the construction of forecasts or warning messages; and the dissemination of warning messages.

A range of early warning systems that were either operative or were in the process of being established were identified by Walker (1989), including the FAO Global Information and Early Warning System (GIEWS), established in 1975 following the 1974 World Food Conference; the USAID-funded Famine Early Warning System (FEWS), established in 1985; the UNICEF Food and Nutritional Surveillance Programme, set up in 1987; Oxfam's Monitoring and Early Warning System (MEWS); the drought monitoring programme of the Sudanese Red Crescent; as well as embryonic systems established by the Save the Children Fund (SCF) in Mali and Ethiopia, the Belgium-based Association Européenne pour le Développement et la Santé (AEDES) in Mali and Chad, and the Geneva-based NGO Ad-Hoc Committee on Disasters. However, as Guha-Sapir and Salih (1995: 10) note, 'it is one thing to collect information and spread warning messages, [but] it is another thing altogether to put into action the policy recommendations made available by this technique.' Walker argues that any organisation providing an early warning system needs to have a clear idea of what they wish to achieve, and that, in particular, the dissemination of a warning message should trigger some predetermined action or set of actions. The difficulty for those wishing to use EWS information for EIAs in refugee emergencies is that of filtering out relevant information from systems that have been increasingly refined to serve specific – and rather different – purposes.

Perhaps the key point in establishing a procedure for EIA in refugee emergencies is not so much the gathering of information on potential environmental impacts – a process likely to produce voluminous documents that emergency staff have little time to read, let alone act upon – but rather to develop a system to prioritise environmental impacts which should be the focus of policy. One proposal for establishing such a system has been put forward by Kelly (1996) for 'disasters' in general, which builds from a theoretical framework rooted in the 'basic needs' of the affected population, as well as a list of factors that Kelly sees as influencing the scale of environmental impacts. The result is two generic 'ratings tables' in which first basic needs and then 'impact factors' are rated on a scale of 1–10 as to their importance. The idea is to provide a non-location-specific method for evaluating the relative urgency of dealing with different needs (defined as water, energy, personal protection and health), as well as the relative importance of different impacts (number of displaced, duration, density, etc.). This approach has a certain amount in common with the procedure of 'multi-criteria analysis' (MCA), in which possible interventions are rated according to the extent to which they address different goals, rather than the single goal of economic efficiency (Van Pelt, 1993; De Graaf, 1997). This allows for the possibility of discussion between actors as to which goals (e.g. economic, ecological, social goals) should be prioritised in any action to mitigate negative impacts – a discussion that is important given the wide potential variations in the perception of environmental problems between refugees, host communities and assistance agencies (Wilson, 1995). There are also interesting parallels to be drawn with 'strategic environmental assessment' (SEA), in which again, alternative approaches of policies, plans or programmes are compared and evaluated (Thérivel and Partidário, 1996). Both of these approaches leave open the option of consultation with potential stakeholders in projects and mitigative actions, both prior to and during implementation.

Guidelines for environmental management

The previous two sections have focused on mechanisms for identifying and predicting environmental impacts in refugee-affected areas, but an additional and essential part of any 'toolkit' for environmental management involves the measures that might be taken to mitigate or avoid any negative environmental impacts that are identified. Indeed, there is little point in identifying and measuring environmental impacts unless some policy interventions follow; and it is part of the standard procedure of an EIA that an 'environmental management plan' is an outcome of the process. However, once again, the development of environmental plans is not

Table 3.2 Environmental activities recommended by the UNHCR

General activities	Emergency phase	Care and maintenance phase	Durable solutions phase
• Modify sectoral guidelines to include environment • Promote environmentally friendly procurement • Co-ordinate policy with other UN agencies, host governments and donors • Promote environmentally friendly technologies • Upgrade the environmental database (incl. GIS) • Provide environmental training	• Include environment in contingency planning • Include environmental specialist or 'focal point' in emergency team • Co-ordinate action with other agencies • Conduct post-emergency assessment	• Appoint environmental co-ordinator or 'focal point' • Prepare environmental master plan (EMP) or environmental action plan (EAP) • Establish environmental task force (ETF) • Include environment in budget planning	• Consider rehabilitation in areas refugees have left • Follow other UN agencies' environmental guidelines in local settlement • Include short-term environmental measures in areas of repatriation

Source: UNHCR (1996).

a straightforward process, and it is to the broad principles on which policy interventions in this area might be founded that this section is devoted.

As noted at the beginning of this chapter, the UNHCR has been very active in the last few years in developing environmental policy, with a specific recent development being the publication in 1996 of a set of *Environmental Guidelines* (UNHCR, 1996). These cover pre-planning of refugee emergencies, the emergency phase itself, the 'care and maintenance' phase which follows an emergency, and the period of what the UNHCR terms 'durable solutions' – traditionally local settlement, resettlement and repatriation. As can be seen from Table 3.2, in general terms these 'guidelines' include recommendations to promote environmentally friendly procurement and technologies, upgrade the organisation's GIS database and capacity in remote sensing noted above, provide environmental training for staff, modify existing sectoral guidelines, and co-ordinate all this effort with other UN agencies, host governments and donor institutions. On these general provisions, some progress has been made – notably with the publication of detailed environmental procurement guidelines (Ross, 1997) – although sectoral guidelines and training were still being developed by the UNHCR at the end of 1997, whilst co-ordination of other UN agencies might be seen as a hopeless task.

In addition to these provisions, though, two further key areas are of particular interest here: first, during the emergency phase, to appoint an environmental specialist or focal point, who has the task of conducting a 'rapid environmental assessment'; and second, during the care and maintenance phase, to establish an 'environmental master plan' and/or 'environmental action plan', overseen by an environmental co-ordinator, and an environmental task force. An environmental master plan (EMP) is conceived by the UNHCR as being the basis for establishing an overall strategy for the environment, usually at a country level and over a five-year period. Initiated by the UNHCR, the development of an EMP is expected to include environmental analysis akin to an EIA, before setting out a 'broad plan of action containing representative projects that best mitigate the most significant negative environmental impacts resulting from the refugee presence' (UNHCR, 1996: 60). The EMP is then to be operationalised in individual projects through 'environmental action plans' (EAPs), which are expected to cover a sub-set of refugee camps within a country, or a particular environmental sector, such as forestry, and to feature as a component of the annual budget submission of the UNHCR with respect to field operations in the area. Such EAPs are seen as lasting no more than three years, although they could be 'rolled over' if funding allows.

To date, there are few examples of the implementation of an EMP or EAP by the UNHCR on which to base an assessment of the approach. One of the more valuable elements though appears to be the linking of the EAP to the establishment of an environmental task force, whose purpose is to bring together various actors and institutions to discuss environmental matters, and ultimately to draw up or contribute to the EAP. However, an environmental action or master plan is only as good as the analysis of environmental problems allows it to be; in other words, if an environmental impact is ignored, mis-specified or exaggerated, any remedial measures enshrined in a plan will also be misplaced. In addition, although the UNHCR's guidelines provide for an environmental co-ordinator and task force, there is a real danger that the individuals and projects assigned to an 'environmental' heading might be marginalised in mainstream decision-making. For example, there may be a sense that by establishing a task force, environmental concerns are already dealt with; whilst at the same time, there is the feeling that (perhaps rightly), environmental concerns cannot be allowed to dictate policy where this would directly lead to a reduction in the already poor livelihood conditions of refugees.

Two separate ways around this conundrum are sought by the UNHCR in its environmental guidelines. The first is the enumeration of a set of 'basic principles' which underline the guidelines, and are seen as the basis for all environmental activities. Thus such activities are expected to be based on an 'integrated approach', in which environmental activities are integrated into other policy interventions (including financially); the principles of 'prevention before cure' and 'cost effectiveness and net benefit maximisation' are to be adhered to (although to date there is little experience of bringing environmental valuation to bear on refugee decision-making); whilst participation of local and refugee communities is to be sought. In addition, included in the guidelines, parallel to specific 'environmental' activities, is the expectation of a modification of other sectoral guidelines to reflect environmental concerns. The main measures proposed are listed in Table 3.3, and cover a range of specific measures advocated in each sector to minimise the likelihood of environmental damage, and in the case of forestry, to re-establish forest cover where this has been removed. On the whole, the measures proposed by the UNHCR relate to attenuating demand, especially for wood, although the promotion of low chemical input or organic farming would appear to imply an increased demand for land under more extensive agricultural systems. There is also a strong role for environmental education, reflecting a belief that changing people's attitudes is crucial to resource protection. This, in turn, fits in with the notion of refugees as 'exceptional resource degraders' outlined in Chapter 1, whereby education is needed to reach people who otherwise have no commitment to the long-term sustainability of their natural resource use.

Table 3.3 Sectoral activities of environmental importance promoted by the UNHCR

Sector	Activities
Supplies and logistics	• Adequate supplies to avoid burden on local environment
	• Environmentally friendly procurement
	• Avoid excessive transportation
	• Reduce packaging and remove waste
Physical planning	• Respect carrying capacity of site
	• Avoid environmentally sensitive areas
	• Incorporate specific environmental measures in site planning
	• Protect existing vegetative cover where possible
	• Use environmentally benign or sustainably gathered shelter materials
Water	• Protection of water supply areas from pollution
	• Ensure proper control of chemicals and waste water
Sanitation	• Establish system for disposal of human excreta
	• Control waste water and/or establish drainage
	• Proper management of garbage
	• Design camp to control dust
	• Insect and rodent control measures
Health	• Identify potential environmental impacts on health
	• Train staff and refugee community

Table 3.3   (Cont'd)

Sector	Activities
Food	• Provide food that requires less or no fuel to prepare • Promote energy-efficient stoves, cooking utensils, cooking methods • Promote community-based grinding of grains • Select foods with low transport, handling and packaging needs
Domestic energy	• Lower demand by promoting efficient energy use (first choice) • Supply alternative fuels (second choice) • Sustainable provision of woodfuel (third choice)
Forestry	• Assess forest resources and draw up plan • Preventative and mitigative action, including proper site planning, forest protection, and controlled wood extraction • Reforestation and afforestation • Monitoring
Agriculture	• Ensure secure access to sufficient agricultural land • Implement soil and land use surveys • Support sustainable farming methods and technologies, using appropriate inputs and extension services
Livestock	• Establish carrying capacity • Consider sale of livestock, slaughtering, negotiations for access to grazing land, relocation of livestock and/or supplementary feeding • Promote animal health and livestock-related public health measures
Community services	• Include environmental concerns in any participatory mechanisms established • Identify refugees skilled in environmental matters • Promote environmental information, awareness and training • Facilitate interaction and conflict resolution between refugees and local communities • Assist in mobilisation of refuge labour for environmental projects
Education	• Develop environmental teaching materials • Train teachers on environment
Income generation	• Promote income-generation activities which contribute to sound environmental management • Identify and discourage environmentally harmful income-generating activities • Design and implement environmentally sound vocational training

Source: UNHCR (1996).

The UNHCR's guidelines represent a step forward in promoting environmental awareness, not only amongst refugees, but perhaps more importantly, among international agencies which provide assistance to refugees. However, some problems remain both with the specific guidelines adopted by the UNHCR and with the general principles involved in establishing environmental plans in refugee situations. Taking a general point first, according to the UNHCR guidelines, the appointment of an environmental co-ordinator and the development of an EMP or EAP in an area is not expected to occur in all refugee emergencies; rather these are measures to be taken either where refugees 'may be sited near sensitive or valuable ecosystems', or where 'population density . . . is likely to exceed the local carrying capacity . . . or lead to serious depletion of natural resources' (UNHCR, 1996: 17–18). But as was noted in the previous chapter, this begs the question of what is a sensitive or valuable ecosystem, the local carrying capacity, or the likelihood of serious depletion of natural resources, when each of these concepts or processes is complex and open to debate. In practice, each is unlikely to be known in detail for a particular area prior to the appointment of an environmental specialist whose job is to establish how important they are. In turn, given that the environmental specialist's own job depends on the existence of such threats to the natural environment, it would be surprising if her/his conclusions were that environmental considerations were not serious!

For example, the notion of a 'sensitive or valuable ecosystem' is commonly equated with the concept of biodiversity; or the idea that preventing the irrevocable loss of habitats and species is of key importance to the world's future. Concern about biodiversity reflects the need to maintain the diversity of the genetic pool, not to mention the potential commercial value of as yet undiscovered plant extracts, etc., in the pharmaceutical and other industries (Wilson and Peters, 1988). The most salient criteria for setting priorities in the conservation of biodiversity can be seen as the distinctiveness of a species or ecosystem, its utility (either globally or locally, in the present and in the future), and the extent of any threat to its existence. However, more practically, refugee assistance agencies must almost inevitably follow existing designations of valuable ecosystems – the 8500 or more protected areas in 169 countries which cover 5 per cent of the world's land surface (Pimbert, 1997).

Such a reliance on existing designations of protected areas as sensitive ecosystems begs a number of questions about the reliability of the science and politics that led to their designation in the first place. For example, Adams (1990) has pointed out the close relationship between the early development of national parks in Africa and colonial hunting lobbies – hardly a designation based on scientific principles of conserving endangered species, and also incidentally a process that resulted in considerable conflict with local populations. Such conflict has remained to the present day; as Pimbert (1997: 420) notes:

> The dominant ideology underpinning [protected areas] . . . has been that people are bad for natural resources. Policies and practices have, therefore, sought to exclude people and so discourage all forms of local

participation. . . . Social conflicts have grown in and around protected areas, and conservation goals themselves have been threatened.

Meanwhile, the dual role of many protected areas in the present day as both repositories of species or habitats which are 'irreplaceable', and as either tourist destinations or source areas for tropical timber or pharmaceutical products, highlights the ambiguity of making a 'scientific' judgement about areas that are worth protecting. In one sense, the development of an environmental management plan represents a sensible precaution if refugees are located close to such areas where conflict between environmental and human (economic) priorities is high; but the limitations of science and conservation goals in providing a framework for such a plan need to be recognised. Alternatively, if the objective of such a plan is simply to keep refugees away from protected areas, the result may be to intensify conflicts. One could argue though that modern refugee policy would simply be fitting into a long history of outside intervention in this respect.

Finally, a number of the specific measures spelt out by the UNHCR remain open to debate. For example, although couched in conditional language, there is mention in the UNHCR's guidelines of the promotion of fuel-saving stoves, controlled wood extraction and reforestation, all of which are policies questioned in Chapter 4. Meanwhile, as in the stage of carrying out an EIA, in citing principles for the development of a master or action plan, there is no sense of how such a plan should prioritise different measures, at least not in specific terms. The result is that certain types of measures become prioritised because of their familiarity, or because of existing capacity to implement them, rather than because they are the most appropriate measures in the circumstances.

Participatory environmental appraisal

Perhaps the most significant feature of the UNHCR's guidelines is the focus on underlying principles – and in particular the principle that environmental actions work best when carried out in a participatory manner with local and refugee communities. If this point is accepted, there is also a strong argument that participation should not be limited to the implementation of projects already identified by agencies after the carrying out of an environmental assessment; but that rather, participation should be sought in the process of assessment itself. As Hoerz et al. (1996: 48) note:

> So far, EIAs are carried out by short-term consultants, contracted by UNHCR or implementing agencies. Participation of refugees, local population and host government staff is often reduced to question and answer sessions with little time for representatives to engage in a consultative and reflective process with their own communities.

In other words, there is a risk in stressing the virtues of 'participation', that this is conceived of simply as participating in environmental activities that are already decided upon – hardly a radical step, particularly if those 'participating' are paid, or given other incentives to take part in those activities. In contrast, much of the growing literature on participatory approaches to development stresses the importance of local participation in *decision-making*, starting at the point of problem identification and analysis. This, for example, is the thrust of one of the most significant trends in participatory thinking to date, namely the development of 'participatory rural appraisal' or PRA (Chambers, 1997).

PRA, and a number of other participatory techniques, have their origin in the development literature, and in particular in frustration with both the time-consuming and overly technocratic and 'scientific' nature of academic study on the one hand, and the shallowness of many consultants' reports on the other. Drawing in part on the earlier approach of 'rapid rural appraisal' (RRA), such techniques might be seen as highly relevant to refugee situations, adhering to the need to provide a relatively rapid assessment of the problems and potentialities of a situation (Chambers, 1982). Certainly PRA is some way off the approach of implementing a standardised format for EIA, leading to standard packages of environmental intervention. In contrast, PRA seeks to mobilise community knowledge to the benefit of communities, allowing them to develop their own analysis of the situation, as well as actions that could take the process of 'bottom-up' development forward. For example, a PRA exercise conducted in the refugee-affected village of Nonah in the Forest Region of Guinea, where encroachment of refugees on a nearby forest reserve had been identified by agency staff as a major problem, highlighted villagers' own conception of the 'problem' as alienation of the reserve from their traditional land area, and ongoing confusion over the nature of land rights in general (Diallo et al., 1995). Their solutions focused not on control of the refugees' activities, but on revision and clarification of the land law.

Some agencies have pressed forward with the development of participatory techniques for environmental appraisal (Pretty and Guijt, 1992), although there remain few publically accessible documents describing the implementation of this type of approach in a refugee situation. Wandira and Hoerz (1997) have carried out a 'participatory environmental appraisal and planning' (PEAP) exercise for GTZ in two refugee settlements in northern Uganda, but found some difficulties with the approach. In particular, they report (unsurprisingly) that a lack of security in the area caused severe difficulties for an exercise that was supposed to last ten days; whilst they also note that 'certain key tools of a PRA will not work with refugees during the first months of settlement. Before the settlement takes place, they (i.e. the refugees) have no idea about the area and even during the first six months their knowledge of the area will be limited' (Wandira and Hoerz, 1997: 9). Of course, this does not preclude developing some sort of participatory exercise focused on local populations, or the conduct of such an exercise after the emergency phase has passed and the refugees have become more settled and familiar with the area. Another alternative approach suggested by Wandira and Hoerz is to conduct the appraisal in fortnightly sessions, each lasting only one to two days.

Another example of a participatory assessment of a refugee situation, in which environmental issues were highlighted but not the primary focus, is also provided by northern Uganda, where Oxfam is the main implementing agency for the 'Ikafe' programme to assist 55 000 Sudanese refugees in Arua District. Here, the agency implemented a participatory review of their programme a year and a half into its operation (Neefjes and David, 1996). Whilst not a specifically environmental review, their report is interesting in its reporting of the perspective of different stakeholders on the nature of the refugees' and the programme's 'impact'. Thus whilst Oxfam staff highlighted 'environment and health' as a major issue, citing the establishment of tree nurseries as an appropriate response which is 'often barely considered in refugee programmes until after the damage is done' (Neefjes and David, 1996: 16), neither national Ugandans nor refugees mentioned 'environment' as an issue, although 'changes in land use' were remarked upon by Ugandans.

In some respects, the development of participatory environmental appraisal processes can be seen to complement the type of developments in monitoring, assessment and planning outlined in this chapter. For example, Tabor and Hutchinson (1994) urge the use of indigenous knowledge and classification systems as the basis for establishing GIS databases, and to complement (rather than replace) conventional land surveys and use of remote sensing. They cite the example of the Senegal River valley, where soils on different land types are distinguished by local farmers even though they are similar in composition, and therefore might not be differentiated using conventional techniques. They also note that maps in digital form are easily and inexpensively archived, updated and disseminated back to the communities that helped to compile them. Similarly, GIS can represent an important technique for verifying the utility of indigenous classification systems, particularly through their ability to show spatial correlations. For example, Lawas and Luning (1996) show an association between elevation and cropping intensity in the Philippines, confirming the relevance of an indigenous classification system which identified elevation as a key variable.

A participatory approach is also one way of ensuring a gendered approach to analysis of environmental issues, since perceptions and experiences of environmental problems are likely to be highly differentiated between men and women. As in non-refugee situations, women are likely to be the primary users of local natural resources such as biomass, soil and water, whilst their control over these resources may well be limited by patriarchal structures that ensure the dominance of men in decision-making (Picard, 1996). Without overemphasising the ability of a participatory approach to tackle negative elements of the gender division of labour or of discrimination against women in decision-making, and without essentialising the role of women with respect to the environment (Jackson, 1993), a participatory approach does at least in principle signal the intention to include all stakeholders in the process of addressing environmental problems (Frischmuth, 1997). This may be particularly relevant in a refugee situation, where there may be a relatively high proportion of women in the population, and where traditional gender roles may be in something of a state of flux (Forbes Martin, 1991). And even if, as Daley (1991) noted for the Burundian community in Tanzania in the late 1980s, there is a balanced sex ratio, gender may still represent an important factor in explaining differential access to natural resources.

However, participatory approaches to environmental impact assessment are not without their problems. First, it is clear from the examples of northern Uganda cited above that where participatory appraisal techniques have been used, environmental issues often are not necessarily highlighted by local populations or refugees. Rather, those invited to 'participate' in planning may perceive more pressing problems concerning access to land, employment, and social facilities such as clinics and schools. This presents a clear difficulty: if the purpose of the review is to identify appropriate environmental interventions, but 'stakeholders' prioritise other issues, how should the external agencies involved respond? This is especially problematic if external 'expertise' is in environmental issues rather than, say, employment generation or the provision of health or education services. The problem is also particularly acute if, as is often the case, the issues prioritised by villagers are fundamentally political rather than technical, and therefore largely outside the control of most or all external assistance agencies. It is little better if environmental issues *are* prioritised by stakeholders, but on the grounds that they understand this to be the only issue that external facilitators wish to discuss.

Secondly, of great significance for refugee situations is the question of which community – or indeed which sections of the community – should carry out, or are likely in practice to carry out, the 'participatory' appraisal. Logically, such a process might be expected to focus on the host community, since it is they who are most likely to depend on local natural resources in the long term, and might have a greater knowledge of the local environment, depending on where the refugees are from. But without participation of the refugee community, the scene may be set for an exclusionary definition of environmental priorities which focuses on 'blaming' refugees rather than seeking common solutions. At the same time, different sections of the host community are likely to be affected in different ways by the presence of refugees, affecting their definition of any problems before the appraisal even starts. Although such difficulties are clearly not insurmountable, the process of bringing communities together to the point at which a participatory environmental appraisal is feasible is likely to be too slow for decision-makers in refugee assistance agencies (and donor and host governments). Of course, the process of bringing the two communities together may well be beneficial in itself.

Finally, as Seldman and Seldman (1995) note, both community-level resource analysis (via PRA) and another increasingly popular 'cure-all' for development problems – democratisation – have failed to stem the hunger and disease in Africa to the extent that many have hoped or expected. They argue that what is needed is a more 'all-encompassing democratic participatory law-making process' (Seldman and Seldman, 1995: 10). In part, this reflects the likely impotence of PRA analyses in situations where these are not linked to the means to implement solutions to problems – where, for example, finance for required initiatives is lacking, or where fundamental legal or political change is necessary but blocked by government or vested interests. For Seldman and Seldman,

PRA needs to do more than simply analyse or 'appraise' to represent a solution; it needs to be capable of identifying the nature of the problem; formulating and proving hypotheses as to its causes; assessing the social costs and benefits of alternative laws to reshape behaviour; and monitoring and evaluating the impact of those laws once implemented. However, all of this might be seen as too much to expect in the often fraught and politically tense context of a major refugee influx – an issue which forms the focus of the chapters that follow.

Conclusion

This chapter has reviewed existing and potential 'toolkits' available to planners in refugee situations, focusing on options for monitoring, assessing and planning interventions in the environmental field. What has characterised the discussion throughout is a focus on actions of a 'generic' nature that can generate principles to be applied to any and all refugee situations – as distinct from discussion of specific projects and interventions, which are considered in more detail in Chapter 4. However, focusing discussion at this level does raise a wider question about general policy – namely whether the aim should be to produce a 'menu' of possible projects, i.e. a genuine 'toolkit' of interventions that are known to work; or to produce a set of principles that should be followed in designing environmental activities. Clearly the UNHCR, in its *Environmental Guidelines*, has tended towards the latter, specifying for example that an EIA should be 'rapid' and 'co-ordinated' with other actions, rather than laying down a specific methodology for carrying out an EIA.[3] However, at the same time, the embryonic development of sectoral guidelines, and some at least of the more detailed principles that are to be contained within them, suggest a more prescriptive tone, in which planners are urged to avoid exceeding 'carrying capacity', avoid environmentally sensitive areas, and consider specific measures, notably in the forestry sector.

The evidence presented so far in this book also tends towards a position that would favour the use of environmental 'tools' that are flexible and capable of identifying and dealing with diversity, rather than a prescriptive range of indicators and interventions that can be applied in all circumstances. The complexity of environmental issues in refugee situations does not lend itself to a single, straightforward mode of analysis, or to simple solutions. However, discussion has also focused on 'tools' that can be used which are more generally applied in a 'development' rather than an 'emergency' context, seeking to identify advantages and pitfalls from that context before adopting them as a methodology. It is in this context that the chapters which follow take up the question of the experience of environmental issues, problems and changes in different refugee situations, starting with a review of policy experience across Africa and Asia that none the less deals with location-specific examples, and then moving to examination of a series of particular case studies. In each case, questions are raised not only about the nature of 'impacts' of refugees – in so far as these can be defined or isolated at all – but whether the collection of problems that have arisen in particular refugee-hosting areas can be addressed in policy terms by interventions informed by decades of broader 'development' experience. Perhaps the most important element in this process is to note that in 'development' too, emphasis is arguably shifting away from meta-narratives and simple solutions to more local, specific and bottom-up approaches.

1.   Recently the coverage of topographic maps at 1:50 000 has improved, with such maps available, for example, in Tanzania, Nepal, and parts of Francophone Africa.

2.   SAR sensors were used by the US military during the emergency in eastern Zaire in 1996 to track movements of troops and refugees as camps were broken up by Laurent Kabila's forces.

3.   However, the *Guidelines* do provide a draft Terms of Reference for an Environmental Specialist in Emergencies, which is quite prescriptive in what it suggests that specialists should do.

CHAPTER 4

Environment and development: what is different about emergencies?

The preceding three chapters have considered three questions: why should there be concern over the relationship between forced migration and environmental change; what evidence is there that the two are causally linked; and what should be done in refugee emergencies to identify environmental aspects of the problem? This chapter moves a step further, by reviewing actions that have been taken in emergency situations around the world in response to environmental problems – leaving aside for a moment the question (raised in previous chapters) of whether these problems are adequately identified, quantified, or indeed 'real'. In Chapter 2, three key areas in which environmental problems might be expected to arise were identified, namely impacts on vegetation, soil and water resources. However, of these, impacts on vegetation, and specifically on forest resources, were highlighted as being of primary concern – at least to governments and international agencies – and it is responses to these impacts that form the main focus of this chapter.

The chapter begins with a discussion of attempts to address the 'supply side' in the environmental equation in humanitarian emergencies, specifically through the provision of fuel (and to a lesser extent construction materials) to refugees and other forced migrants in various situations around the world. Such an approach is unusual, however, routinely being rejected as too expensive. Moreover, where it has been tried, it has not proved unproblematic. The chapter then goes on to consider two major elements of 'demand management' of forest resources – first through the provision of fuel-efficient stoves, which are designed to reduce overall energy consumption, and then through woodland management, including the introduction of controls on cutting of wood, and attempts at reforestation. In each of these cases, a wealth of experience has now been generated, both within the field of refugee assistance and more broadly in development situations. The crucial question addressed here is how far approaches in the field can or should learn from each other, or whether they are genuinely comparable. This is explored with reference to the notion of 'community-based natural resource management' (CBNRM) – now a guiding principle of international and state intervention in the natural resources sector,[1] but one which has had little concrete influence to date in refugee situations or refugee assistance agencies.

Supply-side measures: the provision of energy to refugee camps

When refugees arrive in an area, humanitarian intervention is generally focused on what are seen as the four basic commodities essential to sustain life, namely food, water, shelter and adequate health care. From the start, assistance agencies are concerned to provide these necessities to displaced people in a crisis situation; indeed, specialised refugee assistance agencies consider themselves (often justifiably) as very good at the logistics of providing them both quickly and cost-effectively. Arguably there is little be learnt in such situations from the broader practice of 'development' assistance: aid is required fast, and provided it is in line with the aim of saving lives, it is considered appropriate, although this view has been challenged in the past, notably by Harrell-Bond (1986).

One consequence of increasing concern with environmental issues in humanitarian (and development) assistance has been debate as to whether energy should be added to the list of basic necessities that are considered essential for survival. There are a number of reasons why this might be thought necessary, and why logically, fuel might therefore be provided for refugees in the emergency phase. Most obviously, without fuel for cooking, food is usually less nutritious, and often useless because it is indigestible. Similarly, in cold areas (and at higher altitudes in much of the tropics, it is surprisingly cold at night), provision of blankets is of little use if there is no independent means of keeping warm, and excess heat loss may be a major cause of death (Lamont-Gregory et al., 1995). A number of health-related reasons for being concerned with energy supply are beyond the scope of this book (but see Lamont-Gregory, 1995b; Burkholder and Toole, 1995). Rather, concern here focuses on the notion that by providing energy, highly damaging pressure on woodfuel resources of the area hosting the refugees can be reduced. Putting these reasons together, a powerful case can be made that provision of woodfuel should form part of 'normal' assistance to refugees in an emergency – as was concluded by an ECHO-sponsored workshop on priority issues in humanitarian aid in 1995: 'Energy needs should be given the same status and priority as other basic needs such as food, water, health and shelter; relevant agencies should have responsibility that fuel needs are met with minimum negative impact on the environment' (ECHO/CRED, 1995: 89). Of course, a major reason why energy is not generally included as a 'basic' commodity to be supplied to refugees is because of the balance of costs – to supply energy is expensive, whereas the true cost of local supply of woodfuel, in terms of resource depletion, may only be realised in the long-term, and in any case is likely to be borne by the government and communities of the refugee-hosting area. Moreover, unlike food, or even shelter materials in the form of blankets and plastic sheeting, energy supplies are not immediately obtainable from international donors, since there are few energy surpluses that countries wish to offload onto Third World communities, in order to protect their own markets. None the less, there are some examples of the provision of energy to refugee populations, and these are briefly reviewed below.

Perhaps the most substantial programme to supply energy to refugees to meet cooking and heating requirements in recent years was the supply of kerosene to Afghan refugees in the North West Frontier Province of Pakistan from 1979 onwards. Largely funded by the oil-rich states of the Gulf, which supplied the kerosene free of charge, the programme was not specifically motivated by concern about potential environmental problems, but rather by a match between the immediate need of refugees for fuel and the willingness of particular donors – notably Saudi Arabia – to supply aid in the form of oil. Moreover, in this particular situation, cost was hardly a problem, as the geopolitical priorities of the USA to support refugees from the last great battle of the Cold War, and the Arab oil states to support fellow Muslims fighting against communism, coincided to make the Afghans amongst the most heavily subsidised refugees in the world (Findlay, 1993). Another, rather smaller project to supply kerosene to refugees in Nepal does, however, provide an explicit (and reasonably positive) example of the development of such a project directly in response to concerns over environmental degradation. Since 1992, the UNHCR has provided between 3.2 and 3.6 million litres of kerosene each year to Bhutanese refugees living in Jhapa and Morang Districts of Nepal,[2] after the demand for woodfuel from local forests became an issue of conflict between refugees, local Nepalese communities and the Nepali government (Owen et al., 1998a).

An unpublished review of the Nepali experience by Owen et al. (1998a) provides an interesting example of how, under certain conditions, organised fuel supply can be a viable option to reduce pressure on local energy (and hence forest) resources. In this case, the total cost of the project was estimated to be significant, but not overwhelming: at a cost of between $500 000 to $650 000 per year for fuel, and a further $25 000 to $40 000 for stoves, this worked out at approximately $7 per refugee per year. Although it is not possible to compare this with the 'costs saved' in terms of environmental degradation avoided, Owen et al. suggest that the project is well received by both local communities and the Nepali government (not to mention the refugees themselves), such that it would be difficult to suspend from a political and social point of view. A number of aspects of the Nepali experience reviewed by Owen et al. are important to highlight. First, the fact that kerosene is heavily subsidised in Nepal means the resale value is low, and hence relatively little of the kerosene is resold. This also means the fuel is relatively cheap for the UNHCR to purchase (at 18¢ per litre), although the implication is that the Nepali government is helping to subsidise the cost of the whole operation. None the less, there is strong political support for an operation that externalises at least some of the cost of fuel provision. Second, despite fears expressed elsewhere that kerosene is a difficult fuel to use for those not accustomed to using it, and carries a major fire risk in handling and distribution, the perception of the refugees in Nepal appears to be that it is a higher-quality fuel than firewood that is cleaner and faster for many cooking tasks. Third, and perhaps most important, costs have been kept down, and acceptance of the project enhanced, by a participatory approach to implementation of the project. Thus representatives from the Bhutanese Refugee Women's Association (RWA) were selected to run trials of alternative cooking stoves at the start of the project, whilst distribution from 15 000-litre underground storage tanks is achieved by refugee committees, with the implementing agency, the Nepal Red Cross Society (NRCS) playing only a monitoring role. An important result, according to Owen et al., has been transparency and trust in the running of the project.

However, such a positive experience is not universally apparent in projects designed to supply energy to refugees, and this is particularly the case where wood, rather than kerosene, has been the fuel supplied. Perhaps the most significant example of woodfuel supply to refugees is that of projects developed in Tanzania and Zaire during the recent Rwandan emergency, which is discussed in more detail in Chapter 5. Although at the time, this initiative to supply firewood to the Rwandan camps was seen as innovatory, in practice, there is a history of this kind of intervention in other refugee situations. Prior to the 1994 emergency, woodfuel supply programmes had been attempted in Burundi (Owen and Muchiri, 1994) and Rwanda itself, whilst an earlier and significant example was provided by the mass influx of Mozambican refugees to Malawi in the mid to late 1980s. In Malawi, the presence of around one million refugees in an already quite densely populated country was seen as placing a severe burden on local woodfuel supplies. The Malawi government was especially concerned about encroachment of refugees into commercial state forest plantations, notably in Dedza district, which received one of the highest concentrations of refugees, as well as having significant areas of forest planted during and since the colonial era (Wilson et al., 1989). This concern with the environmental impact of deforestation led in 1988 to funding by the UNDP of a $1 million project to supply woodfuel to refugees in Dedza district, involving purchase of wood thinnings from some of the same plantations that refugees had been using themselves. However, the project was abandoned after only a year, as local supplies of wood were deemed insufficient, and the Forestry Department suggested supplying wood from the mature Vipya plantation some 400 km to the north.

A current example of a woodfuel supply project is provided by the Kakuma refugee camp for Sudanese refugees in northern Kenya. Here, refugees were settled in a dry, marginal environment in which it appears to have been assumed by several actors that local supply of biomass for cooking would not be possible. Thus a report by Owen et al. (1998b) states that the local Turkana population traditionally maintained strict rules prohibiting the cutting of trees, because of their importance in providing fodder for livestock. On arrival, refugees were told of this prohibition, and found themselves harassed by local people when they tried to collect fuel and building materials. In response, the UNHCR and its implementing partner the Lutheran World Federation (LWF) organised a 'lifeline' supply initially of charcoal, and then of charcoal and firewood to the camp, meeting up to 67 per cent of household demand at its peak. One of the interesting features of this project is the balance between charcoal and woodfuel, and linked to this, between local and external supply. Thus charcoal was initially supplied by the LWF because it was assumed that the local area was too dry to supply the required woodfuel, and transport over long distances would be more cost-effective with charcoal (even though 50–90 per cent of the calorific value of wood can be lost in conversion to charcoal). No studies had been undertaken of the potential for local rangelands to supply wood on a sustainable basis.

As the LWF began to purchase charcoal, a number of developments occurred. First, local Turkana pastoralists responded by producing charcoal to meet demand; in turn, as deforestation set in, charcoal production within the district was banned by the District Commissioner. As a result, the nearest source of supply was over 300 km away, and costs spiralled to over $330 000 each year (or around $10 per refugee), excluding the cost of charcoal-burning stoves. Eventually, faced with these spiralling costs, a political compromise was reached in which there was a switch to local purchase of woodfuel rather than charcoal, with contracts being given to local women's groups. The irony of the situation is highlighted by Owen et al. (1998b), who calculate on the basis of data collected by a government agency, the Kenya Rangeland Ecological Monitoring Unit (KREMU), that right from the start, the area within 50 km of the camp could easily have supplied the needs of refugees for woodfuel through natural annual production of biomass. They also note wryly that the local people must have realised they were missing out on an important income-earning opportunity, since a new District Commissioner promptly reversed his predecessor's ruling, stating that woodfuel now had to be supplied from *within* the District![3]

As with kerosene supply, one of the key aspects of woodfuel supply in refugee situations comes down to the issue of cost. To supply a bulky commodity such as wood over a distance of 400 km in Malawi was clearly costly and ultimately unattractive to external donors. In neighbouring Zimbabwe, Le Breton (1995: 9) describes how with a small budget for woodfuel provision, the quantities delivered to refugee camps 'represented a negligible proportion of the total consumption, and tended to be viewed as an unexpected bonus'. Yet what the Malawi case highlights is a situation in which the value of the woodfuel that was already being taken by refugees had been identified, but not properly priced (Babu, 1995). The fact that the government of Malawi and/or local populations, who were the owners of the forests, had allowed its use initially, meant that when external cash ran out, it was easy for the cost to be shifted back to these owners. In contrast, in Kakuma, woodfuel was highly valued by local people, who restricted refugees' access to it at no or low cost. However, willingness to accept local provision of firewood was quite different when this value was reflected in the price charged to refugees. In general, the cost of woodfuel provision is much less visible in situations where there are a multitude of owners of a forest or woodland, or where forest products are used by local people in their daily lives but are not commercialised or regulated; but this does not alter the fact that use of a resource by refugees that would or could otherwise be used by somebody else always represents a cost to those users whose access is reduced, whether this is in the short or the long term.

In general, the provision of fuel of whatever kind to refugees represents a relatively unusual occurrence, but one which might be expected to bring with it exactly the same problems as the provision of any other kind of emergency aid commodity. In this respect, two issues in particular deserve highlighting, namely the technical and cultural appropriateness of the aid supplied. The case of kerosene in Nepal demonstrates that an unfamiliar fuel may be appropriate as a means of dealing with emergency needs, even if it incurs some cost, but that this still depends both on the efficiency of the technology and on its cultural acceptability. However, typically, other energy solutions promoted in refugee situations have failed on one or other of these counts. For example, Owen and Muchiri (1994) report poor results of the use of solar cookers, 'fireless cookers' and 'candle burners' in Burundi. Indeed, despite considerable interest in solar energy as a source of cooking 'fuel', experience has been disappointing, with Umlas (1996a: 2) concluding that 'while in certain very specific and circumscribed refugee camp situations solar cooking may be beneficial to both refugees and the environment, solar cooking still faces many operational problems'. These problems included 'cultural resistance' in Pakistan, where Afghan refugees did not use the solar cookers provided since they could not open the oven to test food before it was done, whilst the cookers were unable to cook certain foods – notably the traditional *nan* bread. Meanwhile, in the Dadaab camps of Kenya, a type of solar cooker called 'CooKit' tested by the UNHCR was found to be slow, useless on cloudy days (as with solar cookers in general), whilst their cardboard panels were destroyed in the rain! (Umlas, 1996a). Such issues of technological appropriateness and cultural acceptability of aid are not questions that are limited to the humanitarian sector. However, they are arguably magnified in this sector, where time to experiment and reflect on constraints is often lacking, whilst the real or perceived urgency of the situation might encourage an attitude that sees culture as unimportant when lives are at stake.

Demand-side measures: stoves and the 'woodfuel gap'

Having considered approaches that try to increase the supply of fuel from external sources in refugee situations, this section reviews various attempts to tackle fuel-related environmental problems through reducing refugees' demand for wood. Here, more clearly, assistance agencies can look to the experience not only of humanitarian emergencies, but also to the broader literature on development. Thus the notion of a 'woodfuel gap' between demand and supply, and attempts to do something about it by reducing demand, have their roots as far back as the 1974 oil crisis, when a number of observers drew attention to what they perceived as the 'other energy crisis' (Eckholm, 1975). Wood remains the primary source fuel for the vast majority of the world's population (Leach and Mearns, 1988), and as such, pressure on woodfuel resources is a matter of great significance for the livelihoods of many of the world's poorest people. It is also clear that excessive use of wood for fuel can contribute to deforestation and desertification (French, 1986). Thus even outside the context of humanitarian emergencies, a major response to this gap between demand and supply has been to try to reduce demand or consumption of woodfuel – or better, reduce the need to consume large quantities of woodfuel – notably through development of stoves which either use significantly less wood, or which use alternative, and preferably renewable, sources of energy.

It is important to mention from the start that the very notion of a 'woodfuel gap' has been debated in the development literature, with a number of observers questioning whether an excess demand problem exists at all, at least in terms of rural woodfuel use (Dewees, 1989; Mercer and Soussan, 1992). Certainly, the extent of woodfuel pressure on woodland resources depends on a number of factors, including diet, stove technology, cooking habits, preference for open fires or particular kinds of stove, the nature of the wood used, its origin, quality and availability, and availability of alternative fuels (Morgan and Moss, 1981). In addition, wood is not used simply for cooking but fulfils a multitude of other functions, from providing light and heat when burned to providing the material for construction of houses and the manufacture of a range of wood-based products (both locally, and commercially), such that the woodfuel question is not simply one of providing fuel for stoves. As a result, assessing the independent effect of these and other demands on wood is not a simple task. Indeed, doubts about the theoretical relevance of the 'gap theory' and disenchantment with the apparent inability of improved stoves to bring about significant reductions in consumption in operational situations had led, by the end of the 1980s, to most multilateral and bilateral donors abandoning their stove programmes (Crewe, 1997).

It is not the intention of this chapter to launch a major case either for or against 'improved' stoves *per se* – not least because this is a subject that has been dealt with extensively elsewhere (Foley et al., 1984; Gill, 1987; Barnes, 1994; Crewe, 1997). However, two interesting points arise. First, if interest in stoves had declined amongst mainstream development agencies, why have they continued to be promoted and discussed amongst humanitarian agencies? And second, has the experience of stove dissemination in refugee situations either produced different results or have refugee agencies been able to learn and move on from earlier problems identified in the development literature? Taking the first point, it might be argued that the 'difference' of refugee situations has led agencies to ignore some of the negative experiences of stove programmes elsewhere, and perhaps with some justification. After all, even if the introduction of fuel-saving stoves fails to address underlying causes of woodfuel shortages and deforestation in 'normal' circumstances, there might be an argument that *any* reduction in woodfuel use is valuable in a refugee situation, and can limit temporary and exceptional use of natural resources. Investment

Figure 4.1 An 'improved' mud stove in use in Benaco camp, Tanzania
Source: (photo: Richard Black)

in a technology that can be seen as reducing women's workloads (Nyoni, 1992) could also be seen as justifiable in refugee situations, where women often see their responsibilities and workloads increase (Forbes Martin, 1991). None the less, investment in a technology that demonstrably does not 'work' would be foolish and inappropriate, whatever the *intended* ecological or social benefits.

In practice, there are many examples of the introduction of energy-saving stoves in refugee situations, and unsurprisingly, these have met with a wide range of levels of success (Figure 4.1). One of the more successful appears to have been the initiative of the Fuelwood Crisis Consortium (FCC) to introduce such stoves to Mozambican refugee camps in Zimbabwe from 1989 to 1992. Le Breton (1992) describes how the consortium was established in response to complaints from elderly women refugees about the distance they were having to travel to collect firewood; and although their request was for woodfuel to be supplied to them (which it subsequently was, on a limited basis), this request led on to a much broader environmental programme based on provision of 17 000 improved 'Tsotso' wood-burning stoves within a population of up to 90 000 refugees, tree-planting activities and an environmental awareness campaign. Funded by a range of participating NGOs as well as by the UNHCR, the consortium also sought to play a co-ordinating role on environmental issues between different agencies, as well as between those agencies and various relevant government departments.

It is interesting to examine the experience of the FCC in some detail for the insights it can provide not only on what makes a (reasonably) successful energy programme in a refugee situation, but also on what was unusual about that situation, and on how the FCC learned as an organisation from non-emergency situations.[4] First, the organisation was clear from the start in Zimbabwe that the main cause of deforestation around refugee camps was the cutting of trees for fuel, rather than any other source of demand. For example, cultivation of land by refugees was prohibited, and pressure for new plots by the small local population was limited, eliminating what is often a primary cause of deforestation; whilst even building material for refugee houses was supplied from commercial timber plantations by the UNHCR rather than being sourced locally. Second, the option of promoting fuel-saving stoves was considered only after other options – including provision of fuels such as coal, charcoal and paraffin – had been costed and compared with the cost of provision of stoves. But perhaps more importantly, those developing the consortium drew explicitly on the lessons of stove programmes elsewhere, ensuring that a stove design was chosen only after field testing by refugees themselves (which also provided more realistic assessments of the wood that could be saved), and accurate measurement of the rate of woodfuel use by refugees so that the impact of the stove programme could be properly monitored.

Another example of a stove-dissemination programme embedded within a wider environmental project in a refugee situation is the 'RESCUE' (Rational Energy Supply, Conservation, Utilisation and Education) project implemented since 1994 in the Dadaab camps for Somali refugees in Kenya by GTZ. An 'innovative' feature of this programme, flagged by one of the principal agency workers, is that a range of stoves are offered to refugees for either cash or work on environmental activities, rather than being handed out free of charge (Hoerz, 1995b). Over three years, RESCUE has involved the dissemination of some 21 000 mud stoves and 29 000 metal and ceramic stoves amongst a refugee population of 120 000 in Dadaab. This approach certainly has some benefits: for example, by providing a choice, and requiring refugees to make some investment in exercising that choice, GTZ has a better idea of the popularity of different types of stove. Such 'consumer testing' may be innovative for refugee situations, but represents an important way in which stove programmes moved on from the impasse of the late 1980s (Crewe, 1997). GTZ also suggests there is a potential for a 'double environmental benefit' if dissemination of the stove also involves work on environmentally beneficial tasks (Hoerz, 1995b).

However, the GTZ programme also demonstrates some of the limitations of a stove-based approach. First, project documents suggest that the approach of simply selling stoves met with little success as refugees did not see them as a priority, given limited cash resources. Meanwhile, after a year of operating the programme in Dadaab, it proved necessary to enlarge the list of items that the refugees could 'earn' through their participation in environmental activities, as the market for wood stoves had become saturated. Indeed, as with the Fuelwood Crisis Consortium in Zimbabwe, the 'success' of the project lies less in a concentration on the provision of stoves than in the combination of stove dissemination with other environmental activities. For example, one important element of the project was to focus on improving the security situation for women, both when collecting woodfuel and in household compounds, through the establishment of 'live fences' of the shrub *Commiphora* spp. (Hoerz, 1996b), whilst the project's mid-term review highlights the establishment of an 'environmental working group' of agency staff and representatives of refugee and local communities as an important output (Hoerz and Kimani, 1996).

Attempts elsewhere to introduce fuel-saving stoves have also met with varied success. Thus whilst a somewhat rosy picture is painted of their potential in refugee situations by Gitonga (1995), a contribution by Ross (1995) to the same volume provides a now familiar note of caution, observing that 'the literature on stoves is peppered with such language as "the stove has the potential to save x of fuel" . . .' but that actual savings are rarely tested in field situations. One such test in the camps of Mauritanian refugees in Senegal showed a slightly higher fuel consumption by those using improved mud stoves than by those using traditional three-stone stoves (Black and Sessay, 1997b). A possible explanation lies in the potential for a 'rebound' of consumption as a result of increased ease of cooking (Dufournaud et al., 1994). Whether or not this is the mechanism at work, such findings highlight the fact that inefficient cooking technology may not be the only or even the main reason for high levels of woodfuel use. Indeed, if it is assumed that the main control on woodfuel use is the amount that households can collect, then the principal impact of an improved stove project may simply be to allow households to cook more, or to keep fires burning for longer, for example for heating and lighting. Of course, such an outcome may not be negative from the user's point of view, although the lack of broader environmental gains may undermine the economic case for investing in the technology. Even in the case of Zimbabwe, there were hopes that the 'success' of Tsotso stoves would be consolidated through their use on return to Mozambique, hopes that were dashed (G. Le Breton, pers. comm.) when the former FCC director visited Mozambique and found refugees using the stoves as buckets!

A number of lessons can be drawn from specifically refugee-related experience with fuel-saving stoves. Perhaps the most obvious is the fact that one of the more successful programmes – that of the Fuelwood Crisis Consortium in Zimbabwe – involved a situation in which refugees' demand for woodfuel clearly placed the main pressure on woodland resources of the region, without the complication of other intervening factors. In contrast, if other pressures exist on these resources, refugees may feel less inclined to make savings in woodfuel use when others are not doing so. It is also important to stress that 'success' of an improved stove programme cannot be measured simply in terms of the number of stoves distributed: this is clearly a waste of time if they remain unused, or if they are not repaired when they become damaged. Rather, 'success' needs to be measured by field monitoring of the results of distributing the stoves; in other words, do the women who are given or buy them like or use them? In short, both the dissemination and monitoring phases can be enormously enhanced by a participatory approach in which there is open and honest communication between agencies and project bene-ficiaries, rather than a 'them' and 'us' atmosphere in which the agencies assume that what they choose is best for the refugees, and, more often than not, the refugees assume precisely the opposite.

On closer analysis, such conclusions can be seen not to be limited in their relevance to refugee situations but are remarkably similar to con-clusions about energy (and other aid) interventions more broadly in the developing world. Even where the exceptional nature of a refugee situ-ation might be thought to justify an exceptional approach, where agencies have tried such an approach, they have generally pulled back from imple-menting it after seeing the results. A case in point is the question of whether, in order to dramatically reduce woodfuel demand, cooking for refugees should be done on an 'institutional' basis with centrally located fuel-saving stoves providing 'canteen' style meals. Such an approach was introduced in the Zimbabwe refugee camps in the 1980s, motivated initially not by a desire to save woodfuel but because in line with the closed nature of these camps (they were surrounded by barbed wire), institutional cooking was seen as reinforcing the temporary appearance of the camps, and en-couraging the refugees to continue to regard Mozambique as their home. However, prior to the establishment of the FCC, such an arrangement had been abandoned since it was seen as undermining the integrity and cohesion of households and the important social and cultural functions that cooking within the household often plays. In spite of the FCC's find-ing that large fuel-efficient stoves used for centralised cooking could save up to 80 per cent of per capita woodfuel consumption (based on trials in remaining school canteens), this social argument against institutionalised cooking was seen as outweighing any benefits from reduced woodfuel use.

Woodland management and reforestation

If some parallels can be seen between the conclusions of agencies pro-moting fuel-saving stoves in refugee situations and those attempting to disseminate such demand-reducing technologies elsewhere in developing countries, examples of cross-fertilisation of knowledge are less evident in a third field of intervention, that of woodland management and reforesta-tion. This is despite the fact that reforestation activities remain the most popular type of activity undertaken by refugee assistance agencies to address environmental concerns (Black, 1994b; see Table 4.1), justified by the desire to 'compensate' host governments and communities for trees cut down by refugees. Three large 'compensatory afforestation' pro-grammes that started in the 1980s, in Somalia, Malawi and Pakistan, provide interesting evidence of some of the potential pitfalls of reforesta-tion in refugee-affected areas, contained within extensive evaluations of particular projects.

In Somalia, several reports from the 1980s are available that, despite their age, make useful reading today. A report by Parker et al. (1987) described how the 'Hiran Refugee Reforestation Project', jointly imple-mented by Care, USAID and the National Range Agency of the Somali

Table 4.1 Environmental projects funded by the UNHCR, 1995–97

Country	Type of environmental activity						
	EAP	Forestry	Wood supply	Other energy supply	Fuel-saving stoves	Environmental education	SWC[1]
Algeria							
Côte d'Ivoire	X						X
DRC		X	X		X	X	
Ethiopia		X				X	X
Guinea		X			(X)		X
India[2]							
Iran		X					X
Kenya	X	X			X	X	X
Liberia							
Nepal		X		X			X
Pakistan		X		X	X	X	X
Senegal							
Sudan		X		X			
Tanzania	X	X	X	X	X	X	
Thailand					X	X	
Uganda	X	X				X	
Zambia		X					

Other countries					
Armenia	Burundi	Bangladesh	Burundi	Malawi	Malawi
Burkina Faso	Malawi	Honduras	Honduras	Zimbabwe	Mexico
Burundi	Rwanda	Mexico	Malawi		Somalia
Cambodia		Somalia	Mexico		
CAR			Rwanda		
Cyprus			Somalia		
Guatemala			Zimbabwe		
Honduras					
Malawi					
Mexico					
PNG					
Sierra Leone					
Somalia					
Vietnam					
Zimbabwe					

Notes: [1] Soil and water conservation.   [2] UNHCR not operational in India.
Source: Compiled by the author from various UNHCR reports.

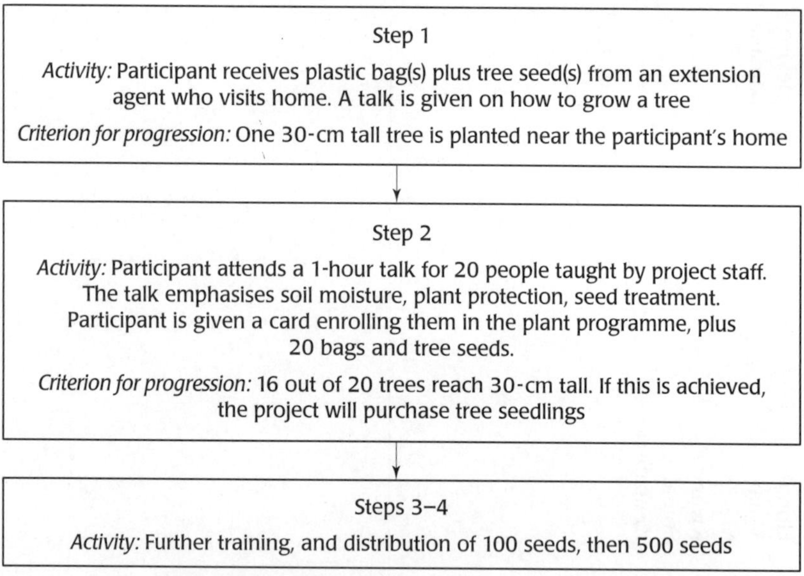

Step 1

Activity: Participant receives plastic bag(s) plus tree seed(s) from an extension agent who visits home. A talk is given on how to grow a tree

Criterion for progression: One 30-cm tall tree is planted near the participant's home

---

Step 2

Activity: Participant attends a 1-hour talk for 20 people taught by project staff. The talk emphasises soil moisture, plant protection, seed treatment. Participant is given a card enrolling them in the plant programme, plus 20 bags and tree seeds.

Criterion for progression: 16 out of 20 trees reach 30-cm tall. If this is achieved, the project will purchase tree seedlings

---

Steps 3–4

Activity: Further training, and distribution of 100 seeds, then 500 seeds

---

Figure 4.2 The 'step' approach to reforestation in Gedo Region, Somalia

Democratic Republic, faced severe difficulties as a result of the selection of exotic tree species and a lack of community participation. After initial failures of the exotics *Leucaena leucocephala* and *Azadirachta indica*, pure stands of *Parkinsonis aculeata* became infected with insect pests. The report concluded somewhat ironically that some of the more convincing environmental rehabilitation actually achieved involved the natural regeneration of native vegetation, after the fencing off and guarding of particular sites. Similar difficulties were observed in a report by Orr (1985) on reforestation in the Gedo region of Somalia with refugees from the Ogaden, although here, a 'step' approach to community participation required verifiable 'results' of initial tree planting by refugees before further training and tree seedlings were distributed (Figure 4.2). In this case, only 15 per cent of the refugees initially enrolling in the programme progressed from 'Stage 1' to 'Stage 2' or, in other words, were 'successful' in keeping a single tree alive in their household compounds, although Leach and Mearns (1988: 176) put a more positive gloss on this project. As Waldron and Hasci (1995: 36) note, such reports could be considered mere ancient history, were it not for the 'predictable reappearance of the same set of problems in the Rwandan refugee tragedy of 1994', which they put down to 'institutional amnesia'.

In contrast to these NGO initiatives in Somalia, in Malawi, a UNHCR-funded project costing over US $5 million between 1988 and 1996 sought to replant a total of 10 000 ha using the state Forestry Department as the main implementing partner (GOM, 1992; Le Breton et al., 1998a). However, the Malawi case highlights a number of additional problems in refugee reforestation. An initial problem faced by the project was a lack of land available for plantations, since in many areas affected by the refugees, land itself was at a high premium. The somewhat ingenious answer was to do the 'replanting' not on land occupied by the refugees (who were in many cases still farming it) but within government forest plantations that had been cut for commercial use. This effectively subsidised the Forestry Department's own plantations, rather than creating additional areas of forest. Second, where trees were planted outside state plantations, there appears to have been little attempt to consult with local communities about species to be planted, or to put in place mechanisms for the subsequent management and ownership of the plantations. In practice, forests on communal land have been handed over to Village Natural Resources Committees (VNRCs), institutions which were recreated in 1994 based on (relatively successful) practice during the colonial era. But according to Sabiti (1996), these committees do not have the training, or the Forestry Department the financial resources, to manage the plantations properly, creating difficulties for their long-term maintenance. Third, as in Somalia nearly a decade earlier, a report by the UK-based Natural Resources Institute at the end of the refugee emergency in 1994 concluded that natural regeneration would have been a more appropriate way to restore the environment in areas with relatively low population densities (NRI, 1994). In areas of high population density, it argued that forest plantations might not be justified given the continuing acute shortage of agricultural land.

In Pakistan and Iran, which between them shared over three million refugees from the conflict in Afghanistan, there were also high profile and high budget schemes based on reforestation activities. In Pakistan, reforestation formed part of a wider 'Income-Generating Project for Refugee-Affected Areas' (IGPRA), funded jointly by ten bilateral donors[5] through the UNHCR and the World Bank to the tune of over $60 million over the period 1984–90, plus a further $22 million spent during 1992–96 as many of the refugees returned to Afghanistan (World Bank, 1996a). Under the first two phases of this project, a total of 43 000 ha of land was reforested, as well as other activities taking place in the fields of watershed management, irrigation, drainage, flood protection, energy and training. The primary aim of the project was to create job opportunities for the refugees and the local population, but this involved targeting environmental activities as one of the major areas in which refugees would be employed, particularly in the third phase of the project from 1992 onwards (Hudson, 1992).

An end-of-project evaluation of the IGPRA project by the World Bank provided a mixed assessment of its success. Thus although the IGPRA provided 24 million work-days of paid employment (two-thirds of it in Phase III targeted at Afghan refugees) and left 'durable assets for the benefit of local populations', the evaluation admitted that repairing environmental damage 'was not a high-priority criterion in selecting sub-projects', as 'repair was impractical so long as refugees and their livestock remained in an area in significant numbers' (World Bank, 1996a: 17). Moreover, the evaluation went on to note that many investments are 'durable *if adequately maintained*' (my italics) – a perennial problem for development projects – whilst stating that maintenance had been a problem in some cases 'since community involvement in project planning and design was not actively sought' (World Bank, 1996a: 18). More worrying were comments that a number of forestry projects were not sustainable because of climate, land tenure and social factors; that there was usually a poor selection of tree species; and that 'IGPRA intervention on *karezes* (underground irrigation channels) in Balochistan may have undermined sustainability by introducing a public sector subsidy into what was a cost-effective private sector activity' (World Bank, 1996a: 18).

An important point here is that whilst the World Bank's survey took as positive the change towards increased vegetation cover, there are a number of aspects of this change that might not be seen in a positive light if the focus is on the 'livelihood environment' of the poorest. Thus the impact on equity was clearly not favourable, one result of IGPRA being the privatisation of former communal grassland and exclusion of poorer members of the society, who previously had 'junior rights' to collect woodfuel and graze animals. In turn, whilst some forestry projects were on private land, distant from refugee concentrations, others were closer to the refugees' 'impact'. However, where the latter was the case, this led to an increased burden on women of collecting fuel and fodder since land was closed for these projects. The report also notes how as time went on, there was a high preponderance of 'very young (school-age) boys and old men' employed by the project, not least because of the absence of many working-age men who had left as migrants to Pakistan's cities, or to the Middle East. Despite these limitations, and the World Bank's cautious conclusion that 'the Pakistan–Afghan IGPRA approach does not provide a blueprint for replication' (World Bank, 1996a: 19), the project has been written about in glowing terms by some within the UNHCR (Durrani, 1996).

Elsewhere too, the problems associated with a tree-planting approach are evident in the available literature, even if the overall conclusions of evaluations have become more positive. In Iran, another 'pioneering' project of environmental rehabilitation was implemented between 1989 and 1995 (Rouchiche, 1996), again with an emphasis on its income-generating aspect, although in this case, the focus was less directly on tree planting activities in the strict sense. Thus the 'South Khorosan Rangeland Rehabilitation and Refugee Income-Generating' project, initiated by the UNHCR, the Iranian government and the International Fund for Agricultural Development (IFAD), sought to 'rehabilitate' some 68 500 ha of rangelands through the planting of trees and shrubs. Meanwhile in the Montebello Lakes National Park in Mexico, the UNHCR responded to concerns over deforestation and loss of biodiversity in one of the country's major forest reserves by implementing a 9200 ha reforestation project (UNHCR, 1991). Both of these projects certainly had some benefits for the host country, and in Iran, the South Khorosan project was no doubt particularly welcome in terms of the employment opportunities generated in a country that remained under a virtual aid embargo led by the United States, and in which assistance to refugees was therefore sparse.[6] Thus a final evaluation report for the UNHCR and the IFAD in Iran argued that the planting of ground cover and shelter belts had had a noticeable impact in reducing dust storms, whilst soil conservation activities had improved water infiltration, although they had not eliminated water run-off (Lachaal et al., 1995). However, a report by Squires (1994) noted that one of the sites 'rehabilitated' by the project had subsequently been earmarked as a site for an 8 million m³ capacity dam, whilst Rouchiche (1996) expressed concern that the sustainability of rehabilitated areas had not been secured. Indeed, despite approving of the overall approach, Rouchiche suggested that 'the positive impacts on range recovery and, to some extent on the carrying capacity, seem to be more imputable to the protection effect provided, following withdrawal of livestock than to the supposed protective action generally attributed by local technicians to the *Haloxylon* and *Atriplex* shrub plantations' (Rouchiche, 1996: 105). In the Mexican example, there was a severe problem of infestation of tree seedlings by the parasite *Denctroctonus* (UNHCR, 1991), requiring at one stage the felling of trees rather than further reforestation.

A number of recurring themes emerge from these and other examples of reforestation or rehabilitation activities in refugee-affected areas that are worthy of mention. First, in each case, despite the expenditure of considerable sums of money, the question remained at the end of projects as to whether allowing natural regeneration might have been a more cost-effective approach to promoting vegetation recovery. Linked to this, approaches have often been top-down, leaving crucial questions unanswered for local communities. Thus, for example, project planning goals frequently specify areas to be planted or numbers of seedlings to be raised without considering the rehabilitation objectives of local communities. Ownership of the trees that are planted is often unclear, or where it is clear, this is often resolved in favour of the state or private individuals rather than local communities (Figure 4.3). Similarly, the intended use of the trees is also often not spelt out, leading to the selection of inappropriate (usually exotic) tree species, suitable primarily for poles or timber rather than the multiple uses that rural people make of forests in much of the developing world. Such species have also been vulnerable to pest attack. In general, the projects lacked baseline data or monitoring that would allow a serious assessment of their efficacy in addressing environmental concerns.

In practice, examples abound of reforestation schemes in refugee situations that have failed largely due to a failure to involve refugees, and especially local communities, in project planning, and where ownership issues are therefore unresolved and motivation of 'recipient' communities low. For example, in Ukwimi refugee settlement in Zambia, Black (1994c) and Lassailly-Jacob (1994b) identified resistance amongst refugees and local Zambians to involvement in an afforestation scheme. What was seen as a 'participatory' approach based on 'voluntary' labour by the aid agency concerned was described as '*chibalo*' by the refugees – the Mozambican term for slave labour! Local Zambians described the scheme as benefiting the administration, but not them (Lassailly-Jacob, 1994b). In neighbouring Malawi, it was noted that tree-planting by children in refugee schools was often done as a punishment handed out by teachers (K. Stevenson, 1997, pers. comm.). Elsewhere, in Guinea, where incentives in the form of cash payment or food have been given to refugees and local people to

Figure 4.3 Privatisation of land in western Côte d'Ivoire
Source: (photo: Richard Black)

plant trees, this is generally, and unsurprisingly, viewed as waged labour rather than 'participation' in an environmental project (Black et al., 1996). Where little or no thought is given to subsequent management and owner-ship of plantations, survival rates of trees are predictably low.

An additional issue that emerges is that of the pressure of the 'project cycle' for refugee situations, which differs in important respects from those in other 'development' situations. Thus the pressure for immediate solutions to environmental (and other) problems comes not so much from the urgency of the refugees' plight but perhaps more from one-year pro-ject cycles which demand that a concrete programme of action is devised and implemented before the end of the financial year. With no guarantee of continued funding in the following year, there is an inevitable bias towards projects that have immediate rather than longer-term objectives, with little or no provision for subsequent management or time to consult on complicated issues such as ownership, access and multiple use.

Community-based natural resource management

Underlying a number of the problems with agency interventions to mit-igate negative environmental impacts in refugee situations, which are highlighted above, is the same issue that was touched upon in Chapter 3 – namely a lack of consultation with refugees and especially with local people over what kind of environmental interventions they would like to see. This is not universally the case, with projects such as the Rescue project in Kenya, and the Fuelwood Crisis Consortium in Zimbabwe, providing examples of considerable time and energy placed on consultation and encouraging participation. Moreover, the UNHCR is increasingly adopting the language of 'participation' in a range of public documents and project activities, in conditions of urgency that are hardly conducive to drawn-out participatory planning exercises. Indeed, perhaps the more important question for refugee assistance agencies is not whether 'community' or refugee participation is desirable, but the extent to which it can realistically be achieved in refugee situations. To what extent is this an area in which 'humanitarians' can legitimately learn from the wider literature on participatory development approaches?

One starting point from which lessons might be learnt is provided by the growing literature on 'community-based natural resource management' (CBNRM). Projects and writing on the CBNRM approach are rooted in the observation that communities frequently have more knowledge of local situations and thus better (or at least complementary) technical expertise than outside 'experts', as well as on the political principle that the interventions of external actors and agencies should be accountable to beneficiary populations as 'stakeholders'. There are now a gamut of techniques, most prominent amongst them 'participatory rural appraisal' or PRA techniques (Chambers, 1997), discussed in Chapter 3, for participatory project planning at a community level. There are also a range of experiences of community-led sustainable development upon which interventions in refugee-affected areas could be modelled.

Without returning to the principles of PRA, a brief analysis of the CBNRM field shows a number of issues that underpin the practical implementation of this kind of approach. These can perhaps be summarised as encompassing the following: ownership (of the resource, and the project) by stakeholders; organisation (i.e. the existence of a community-based structure to manage the project and resolve disputes or debates in its implementation); and economic incentive – in other words, a CBNRM approach is based on the principle that it is in the economic interest of communities to participate in natural resource management activities, otherwise genuine participation will remain low. Perhaps the most well-known examples of this approach are the various schemes to promote sustainable tourism in southern Africa through the involvement of community-based organisations (CBOs). The Communal Areas Management Programme for Indigenous Resources (CAMPFIRE) scheme in Zimbabwe is the best-known (and best-funded) of these, but ADMADE and LIRDP in Zambia, the LIFE programme in Namibia, the Natural Resources Management Programme in Botswana, the Selous Conservation Strategy and the Serengeti Regional Conservation Strategy in Tanzania, and the 'Tchumo Tchato' project in Mozambique are similar examples. Initiated largely at a local level, these schemes have received international acclaim and substantial donor funding, notably from USAID.

Once again, it is not the purpose of this book to describe these schemes in detail, a task that in any case has been done in a relatively voluminous literature elsewhere (Murphree, 1991; Rihoy, 1995; Alexander and McGregor, 1996; Murphree and Metcalfe, 1997). Rather, what is important is to examine the extent to which experiences of the CBNRM approach can be applied to refugee situations; after all, beyond the similarity of the involvement of mass movement, tourism (usually based on game viewing) and refugee situations can hardly be expected to have very much in common. In practice, there are a number of ways in which both positive and 'warning' lessons can be learnt. First, it is clear from schemes such as CAMPFIRE that when communities are taken into the decision-making process in project management (rather than simply being used as paid, or worse, unpaid labour), and where this involvement allows them to identify appropriate ways in which they can benefit from the project (rather than having these benefits specified for them in advance in project documents), management of wildlife can be transformed from a constant battle between communities and the authorities into a mutually beneficial relationship which also reverses the decline of wildlife populations. Since the introduction of the CAMPFIRE approach in Zimbabwe, there has been a decline in poaching, an increase in the elephant population, and real co-operation in natural resource management, where a previous heavy-handed policing approach had failed. The scheme has also brought genuine benefits to rural populations, and especially rural women, for example through the purchase of grinding mills by communities with their share of the profits.

There are a number of difficulties in applying lessons from the specific success of CAMPFIRE – or other CBNRM schemes in the region based on wildlife protection – to the case of natural resource management in refugee situations. Most obviously, these schemes depend for their financial viability on a ready source of revenue – notably the up to $1800 per day that hunters will pay to shoot wild animals (in a controlled, sustainable manner), plus 'trophy fees' – which can be redistributed in part to the community as the 'economic incentive'. Even where wildlife protection is the issue in refugee situations (as in the Virunga National Park in Zaire; see Chapter 5), it is unlikely that stable tourist revenues could be achieved in a situation where armed conflict had already led to mass population movements. More generally, where protection of forest resources is the issue (whether in refugee situations or not), such resources may struggle to be commercially viable at all, let alone provide a surplus for redistribution to the community. None the less, there is emerging interest in the untapped potential wealth derivable from forest products, and their commercialisation by refugees is a feature that has often been remarked upon (*cf.* McGregor et al., 1991; Jacobsen, 1994).

The commercialisation of forest products by refugees has often been viewed by refugee assistance agencies and academic commentators in a relatively negative light, reflecting fears that this process will lead to excessive exploitation and hence degradation of natural resources rather than their protection. Clearly there is a danger of this occurring, and reservations might also be expressed about, for example, Joint Forest

Management (JFM) plans in non-refugee situations. In such schemes, protection of forests is based on the principle of partnership between communities and logging companies, which are seen by some as the second major threat to the forests after farmers themselves (Myers, 1991). None the less, protection of forests through the controlled management of logging rather than its prohibition has become a core part of the approach of a range of important international environmental organisations – not least the World Wide Fund for Nature (WWF) and the International Union for the Conservation of Nature (IUCN). This reflects the reality that the commercialisation of forest products is both inevitable and necessary to meet populations' demands for economic improvement. The challenge becomes one of trying to manage that process to maximise environmental benefits (or at least minimise environmental damage). Examples of JFM schemes are available from other parts of the world, notably India, where JFM has been official government policy since 1988 (Nhira and Matose, 1995), and Nepal.

If the principle of generating an economic incentive for conservation is accepted, the question remains as to whether refugees, and refugee situations, can fit in with the two other requirements for CBNRM projects outlined above, namely ownership and organisational capacity at a community level. In other words, can refugees legitimately be regarded as 'stakeholders'? On the question of ownership, there is perhaps an immediate reply that it is very unusual indeed, at least in the context of most rural areas of the Third World, for refugees to have rights of ownership, whether over land, trees or other natural resources. This might be expected to limit their ability or willingness to participate in schemes for management of natural resources which they do not, and indeed cannot, own. However, this misses the point that not only refugees but also large numbers of the rural poor who are not refugees also lack formal ownership, not least because in many countries most rural land was put formally into state hands, or alienated by wealthy élites or international companies during and after the colonial era. In this sense, it is not so much formal (or private) ownership that is key in natural resource management, but the question of secure access to, and control over, the resources that are to be managed. At their best, CBNRM projects are not only about managing natural resources but also about re-establishing and confirming rights of secure access to and control over them. There is no technical reason why refugees should not be included in this process alongside local communities.

In the absence of a technical barrier, the sixty-four thousand dollar question therefore becomes whether refugees can or should be included in CBNRM schemes designed on the premise of strengthening rights of access and control, and of developing community-level management organisations or institutions. One argument against this was touched upon in Chapter 1 – namely the notion that refugees have a temporary time horizon, and therefore lack the long-term interest to participate in structures that are about long-term sustainability. In reality, both refugees and host communities are faced with having to balance out short- and long-term interests and priorities; both have an immediate incentive to maximise current income; but equally both may have a stake in the longer-term, and may therefore participate in longer-term planning assuming that their short-term survival is not undermined. More crucial perhaps is the attitude of host populations and host governments as to whether *they* wish refugees to be involved in a joint community-based approach to management. Involvement of refugees in what might be termed 'development' projects has never been a popular theme for many host governments, and as was noted in Chapter 1, host government attitudes have hardened in recent years.

Clearly there is some advantage to be derived by host communities from excluding refugees from access to natural resources, or from participation in management schemes from which those communities expect to derive some material benefit. However, willingness to include refugees need not simply derive from altruism. For example, refugees may bring important skills or experiences, as was the case in southern Sudan in the mid-1980s, when agricultural techniques brought by Ugandan refugees were some ways in advance of those of their Sudanese hosts (Harrell-Bond, 1986). Host communities are also likely to be confronted daily with the consequences of exclusion, including thefts of resources and potential for social conflict, which may explain why in many parts of rural Africa, refugees have been able to gain some access to natural resources. If inclusion in natural resource management systems is done on terms determined by the local leaders, this can enhance the power and prestige of those leaders, who effectively gain new 'clients'. These processes are examined in more detail in Chapter 6 for the specific case of Liberian refugees in Guinea.

The problem of host government attitudes towards inclusion of refugees in natural resource management systems is perhaps greater, and presents real difficulties for international agencies that may wish to promote community-based approaches that include refugees. Clearly it is not in the interest of host governments to make CBNRM initiatives dependent on the presence or input of refugees, or to make refugees the primary beneficiaries of such projects. However, the reality at present is that refugees are already the target of significant external aid, and as humanitarian assistance has become more popular, at the expense of development assistance, the extent to which refugees are specifically targeted has arguably grown. In this context, there is some potential for a move back to the principles of the 'relief to development continuum' enunciated in the Second International Conference on Aid to Refugees in Africa (ICARA II) in 1984. This called for humanitarian aid to be linked more to development programmes that would provide long-term benefits for host populations and their governments. Given favourable geopolitical circumstances, the presence of refugees provides a potential lever for host countries to increase flows of foreign aid: the participation of refugees in projects would seem a small price to pay to increase the overall spending of donors. Indeed, it is clear that the failure of ICARA II reflected the lack of interest of donors, rather than unwillingness of host governments to include refugees in development programmes.

Broadening the discussion, there is also increasing awareness in the development-based literature on community-based approaches to natural resource management that communities themselves are heterogeneous, whilst natural resource management often requires collaboration between distinct communities since resource boundaries rarely coincide with community boundaries. It is in this context that the UK's Department for International Development (DFID; formerly the ODA) promotes 'stakeholder participation' rather than simply 'community participation', defining this approach as a 'process whereby all those with an interest (stakeholders) play an active role in decision making and in the consequent activities which affect them' (ODA, 1995: 94). Refugees are clearly 'stakeholders' in the sense that they have an interest in decisions over natural resources in the area in which they live; the key question is whether in practical terms their participation in structures where they could express that interest is viable in the short term, without becoming necessary in the long term. In such circumstances, external intervention could feasibly support that viability, and argue the case with host governments. At the same time, it is important to note that 'participation' itself can have many dimensions, and that the 'social breadth' of participation may itself be important to stimulating the process (de Groot, 1989).

It is necessary to proceed with some caution in promoting a CBNRM approach in refugee situations. Despite scoring some successes, the approach has none the less attracted some criticisms concerning its effectiveness, and especially over questions of equity. For example, Alexander and McGregor (1996) analyse the development of the Gwampa Valley Campfire project in Nkayi and Lupane districts in Zimbabwe, noting the differing attitudes of local councils, which were set to benefit from large financial inputs through the scheme, and local people, who viewed the scheme as a 'backward' initiative over which they had not been consulted, and from which they could see no benefits. In particular, the prospect of eviction of populations from the area to be protected for game animals was seen as a betrayal of the promises of Zimbabwean independence, in which access to land had been such an important issue. Similarly, in the context of tropical rain forest conservation in Cameroon, Burnham (1993: 9) identifies the risk of 'operational carve-ups of zones of influence between the wildlife conservation interests and the logging interests', with each agreeing 'not to address certain awkward issues' concerning community participation and community interests. Hasler (1993) argues that the 'political ecology of scale' suggests that community-level approaches will only work effectively where institutions at higher levels of social organisation have a vested interest in such community participation.

Summarising the literature on CBNRM approaches, Leach et al. (1997a: 1) note that in spite of 'widespread consensus within international development circles that "sustainable development" should be based on local-level solutions derived from community initiatives, there is also a growing perception that the practical implementation of "community-based sustainable development" often falls short of expectations'. They focus on a lack of understanding of how local resource management institutions work as one of the main reasons for this poor record. Perhaps the most important point is that even in non-refugee situations, a community-based approach must face up to limitations on the concepts of 'community' and 'participation'. Thus although it is necessary to search for solutions to what sometimes appear intractable problems, the point is highlighted that these limitations are not unique to refugee situations. Indeed, just as solutions found in 'mainstream' projects for, and literature on, sustainable development can be applied to refugee situations, so too it becomes credible to consider that solutions might be found in refugee-affected areas that themselves have much wider resonance.

1.   The approach also comes under other guises and acronyms, such as 'Local Resource Management' (LRM: van den Breemer and Venema, 1995), or in Francophone countries, 'Gestion des terroirs'.

2.   This represents one litre per person per week for families of up to three persons, plus an additional 0.5 litres per person per week for any additional family members.

3.   A different version of these events is presented by Hoerz (1996a), who states that initially, woodfuel was supplied from local sources using 15 contractors. A subsequent decision by the District Administration to ban collection of dry wood within Turkana District on environmental grounds was said to disappoint locals, who lost an important source of income. Whichever version is correct, the point is that when it is valued and paid for, local supply of woodfuel may be more acceptable to local communities than supply of fuel from outside.

4.   The discussion which follows of both the Fuelwood Crisis Consortium and of the Rescue project is based on both analysis of internal documents of the two organisations and a number of informal interviews with key actors involved.

5.   Donors to the IGPRA were the governments of Canada, Germany, Finland, the Netherlands, Norway, Sweden, Switzerland, the UK and the USA, plus the European Union. The total grant over 12 years was $85.5 million.

6.   The project budget was $20.4 million, although this was never fully funded (Lachaal et al., 1995).

CHAPTER 5

Responding to refugee emergencies: the Rwandan crisis

The refugee crisis of 1994–96, which forced millions of Rwandans into neighbouring Tanzania and what is now the Democratic Republic of the Congo (formerly Zaïre) provides a good starting point for detailed investigation of environmental impacts and environmental policies in refugee emergencies. Between them, the Kagera region of Tanzania and north Kivu represent the first major field operations of the UNHCR in which environmental concerns were explicitly included in policy from an early stage. In addition to being the subject of numerous reports on environmental issues, the Kagera region in particular was the focus of a documentary film, shot in late 1994,[1] on what could be done to address refugee-related environmental change. Based on research conducted at that time, a thorough review of available documentary evidence and discussions with a number of actors involved in environmental operations in the two regions, this chapter identifies the main environmental concerns in areas affected by Rwandan refugees, before discussing a range of policy responses.

The Rwandan crisis and environmental concerns

In terms of size, speed and the severity of its impacts, the refugee crisis that unfolded in Rwanda during 1994 was unprecedented in modern times. The world's media watched as an estimated 170 000 people crossed the Rwanda–Tanzania border on 28 April 1994 (Jaspars, 1994), the largest influx of refugees ever recorded in a single day. Less than three months later, during 14–18 July, as many as 850 000 Rwandans moved across the then Zaïre border into the north Kivu region around Goma, catching the international community and the Zaïrean authorities almost completely unprepared (Borton et al., 1996). With further flows later in the year to Bukavu in south Kivu after the collapse of Opération Turquoise in the south-west of Rwanda, as well as smaller-scale movements to Burundi and Uganda, by the end of 1994 nearly three million Rwandans were living as refugees in neighbouring countries, or internally displaced in

101

Table 5.1 Rwandan refugees and displaced persons in and around Rwanda, December 1993 and December 1994

Host country	1993	1994
Zaire (now DRC)	50 000	1 000 000
Tanzania	50 000	550 000
Burundi	255 000	160 000
Uganda	210 000	5 000
Internally displaced within Rwanda	300 000	1 200 000
Total refugees / displaced	765 000	2 915 000

Source: USCR (1994, 1995b).

camps inside Rwanda (Table 5.1). In response to this massive exodus of refugees, there was also an unprecedented mobilisation of international agencies to deal with the crisis. From three international NGOs present in Goma at the start of the influx, the number rose to 100 within two months, and over 200 by the following year. Meanwhile, many of these NGOs for the first time included 'environmental' issues as a central part of their concern and activities.

A comprehensive evaluation of the response to the Rwanda crisis (the Joint Evaluation of Emergency Assistance to Rwanda) was conducted, one year after the genocide of 1994, with the participation of most of the major agencies and donors involved. This evaluation highlighted both positive and negative lessons from the experience of humanitarian assistance to date (Borton et al., 1996). Although criticised from certain quarters, including from within the UN (Mackintosh, 1996), it drew a contrast between a 'highly impressive' humanitarian operation in Tanzania and a disastrous response in particular in Goma, where an estimated 12 000 refugees died of cholera in the early stages of the crisis (Siddique et al., 1995). In Tanzania, food aid was provided rapidly for an initial planning figure of 250 000 refugees (although this number was eventually exceeded), drawing on stocks of food that were already available in Tanzania's strategic grain reserve at Shinyanga, only a few hours' drive from the refugee-affected region. The rapid response to the crisis was facilitated by the fact that several international agencies, including the UNHCR, were already working in the region, providing assistance to Burundian refugees who had started arriving during the previous year. In contrast, in Goma, contingency planning was for a total of only 50 000 refugees, in spite of considerable warning of a massive influx as events in Rwanda unfolded during May and June. Provision of food aid was hampered by the isolation of the region, such that most food had to be airlifted to Goma in the early stages. Numerous deaths were caused simply by dehydration because of completely inadequate supplies of potable water. This was in part due to the basaltic terrain, in which it was difficult or impossible to dig pit latrines and drill boreholes.

The issue of environmental health problems encountered during the emergency, and in particular the question of whether the deaths that occurred could have been avoided, is dealt with in detail elsewhere (Paquet and van Soest, 1994; van Damme, 1995; Goma Epidemiology Group, 1995). Instead, the main purpose of this chapter is to address the broader question of the 'environmental impact' of the crisis with specific reference to natural resources, and to consider both the magnitude and seriousness of the environmental change that occurred, and the nature of policy responses to it. This is not merely an academic exercise, but one with important implications in terms of drawing conclusions for refugee emergencies elsewhere. It is facilitated by the fact that a number of studies have now been conducted on environmental issues in the region, although few are published or widely available, and the underlying quality of some of the data that underpin them is highly questionable. Indeed, despite some limitations, the Rwanda crisis still represents the first (and perhaps only) refugee emergency in which comprehensive assessments of the environmental impact of refugees have been attempted, and in which the UNHCR and NGOs have deployed environmental co-ordinators and implemented wide-ranging environmental programmes, in order to counter what was seen as the huge potential for environmental damage of the refugees' presence.

The discussion below begins with consideration of the pattern of settlement of the refugees, including why camps were sited where they were, and what might be learnt for future emergency site planning. Attention is then shifted to a number of formal and informal assessments that were made of environmental and environment-related impacts in the Great Lakes region between 1994 and 1996, before considering a range of measures that were proposed or implemented to mitigate negative environmental impacts. In Tanzania, environmental assessments were conducted by the UNHCR (Ketel, 1994a), the aid agency Care (ERM, 1994) and the Tanzanian government (URT, 1995; Kikula and Magabe, 1996), as well as by several other national and international agencies (Green, 1994; Sepp et al., 1995). Similar assessments were also conducted in north and south Kivu (Biswas et al., 1994; Ketel, 1994b; Vincent, 1995) and inside Rwanda itself (ERM, 1995). Taken together, they provide an insight not only into the state of the environment as the crisis unfolded but also into the process of conducting environmental assessments, and the scope and nature of agencies' responses to environmental issues.

Site selection

In many respects, selection of sites for refugee settlement is of paramount importance in determining the nature and severity of environmental impacts resulting from the refugees' presence. Thus in Chapter 2, it was noted that Jacobsen (1994) regards this as the single most important issue affecting the nature of environmental impacts, whilst in Chapter 3, environmental guidelines produced by the UNHCR were reviewed in which it was suggested that to reduce pressure on local environments, attention needs to be paid to the carrying capacity of an area in identifying sites

Table 5.2 Refugee camps for over 50 000 people in and around Rwanda, 1996

District	Camp	Population
Ngara (Tanzania)	Benaco	160 000
	Lumasi	115 000
	Msuhura	80 000
Biharamulo (Tanzania)	Kitali Hill	75 000
North Kivu (DRC)	Kibumba	210 000
	Katale	190 000
	Mugunga	160 000
	Kahindo	80 000
	Lac Vert	55 000
South Kivu (DRC)	Inera	50 000
Total in large camps		1 175 000
Total refugee population		1 880 000
% refugees in large camps		62.5%

Source: compiled from UNHCR reports, rounded to nearest 5 000.

for settlement. The debate on whether refugees should be housed in camps, and if so, how large these should be, has been a high profile one, appearing for example in the pages of the medical journal *The Lancet*, in which van Damme (1995) drew attention to different mortality rates in the camps established in Goma and areas of more dispersed refugee settlement in Guinea (see Chapter 6). In response to this article, van der Borght and Philips (1995) note the advantages of establishing camps for refugees, in terms of service delivery, accountability, identification of individuals, cost-effectiveness and monitoring activities, but accept the health risks of crowded, inappropriately sited camps. In turn, on behalf of the UNHCR, Dualeh (1995: 1369) argued: 'We never choose to house large numbers of refugees in camps. But when this has been necessary, our policy is to urge governments to keep the size of camps to manageable numbers – as a rule of thumb, no more than 10 000 refugees per camp.' The problems of camps are many, and certainly not limited to the environmental sphere. For example, a special issue of the journal *Refugee Participation Network* entitled 'Avoiding Camps' noted a correlation between the size of refugee camps and levels of mortality, whilst suggesting that they undermine refugees' livelihood strategies, generate hostility from host populations, and lead to a loss of autonomy for camp populations (RPN, 1991). Hyndman (1996) has launched a powerful critique of the preference for camps in Kenya, which draws on an equally powerful case made a decade earlier by Harrell-Bond (1986) in southern Sudan. However, it is clear that sometimes, political or practical circumstances mean that the establishment of temporary camps is the only realistic option.

In these circumstances, it is striking that the overwhelming outcome of mass displacement of populations both inside and outside Rwanda in

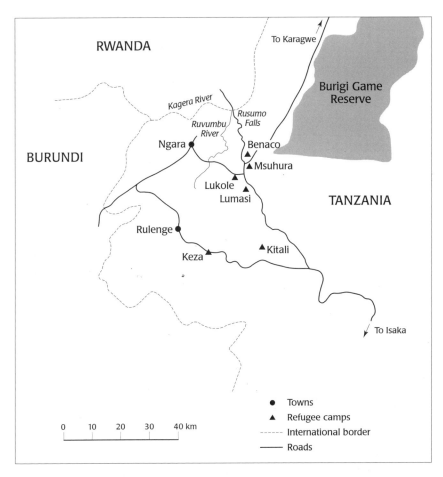

Figure 5.1 Location of refugee camps in Ngara District, Tanzania, 1994–96

1994 was the creation of relatively large camps (Table 5.2), with potentially serious consequences for local environments as well as the affected population. Thus in Tanzania, around 250 000 refugees were initially located in a camp at Benaco in Ngara District, some 12 km from the border crossing to Rwanda at Rusumo Falls (Figure 5.1). Despite the creation of new camps at Lumasi, Msuhura, Keza, Lukole and Kitali Hill to aid in the 'decongestion' of Benaco, a very high concentration of refugees was maintained within a limited area (Figure 5.2). In Goma, camps were initially established for 110 000 at Katale, 135 000 at Kibumba and 125 000 at Mugunga, with two smaller camps emerging later as new sites were sought. The influx into Bukavu, although large, was slower, and whilst initial concentrations built up in Bukavu town, notably at the Alfajiri Seminary/College, which hosted 20 000 refugees during August 1994, by the end of the year, around 300 000 refugees were spread between 25 camps, the largest being Inera

Figure 5.2   Benaco camp, Tanzania, November 1994
Source: (photo: Richard Black)

with just over 50 000 refugees. The existence of such large camps in two of the three main zones hosting the refugees, as well as for displaced people within Rwanda itself, merits some attention as to why, in spite of the increasingly high profile of environmental arguments by 1994, and a range of other arguments against camps, this situation should have come about. Indeed, if the forces leading to the creation of camps are so strong, the question might reasonably be asked as to whether there is any point in drawing up guidelines recommending against the creation of such camps.

   In the case of Benaco, the initial site for the refugee camp had been identified prior to the arrival of the refugees as a potential site for local settlement of some 15 000–20 000 Burundian refugees housed in temporary camps further to the south (Jaspars, 1994). Its main advantage was that it was located adjacent to the main road running from Rusumo Falls (Rwanda) to Isaka, the railhead used by the WFP to offload food aid, in an area with hardly any local population, and next to an artificial lake that had been created by the Italian road construction company that had been working in the area (Figure 5.2). The lake provided an important source of water, although as the population of the camp grew rapidly, it was unable to supply more than 60 per cent of the water required, and even then with a serious drawdown of the lake level. The site was located in an area of *Acacia* and *Brachystegia* woodland, which was felled to clear space for huts and also provide much needed wood for construction and fuel.

   In Goma, more difficulty was encountered in site selection, both because pre-planning was limited, and because of the sheer size of the influx. The

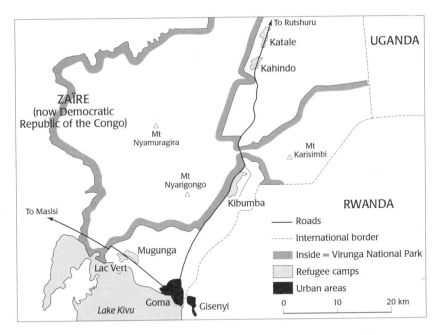

Figure 5.3 Location of refugee camps in north Kivu, Zaire, 1994–96

majority of the refugees entered north Kivu from Gisenyi town to Goma town, although around 300 000 also entered further north, near Kibumba (Figure 5.3). In their evaluation of the initial relief operation, Borton et al. (1996: 35) provide a damming report:

> Those entering Goma were encouraged to keep moving through the town, with most initially being directed on the north axis towards Kibumba and Katale by UNHCR and NGO personnel. Between Goma and Katale, much of the terrain is waterless lava fields, and the distance to the planned (but still unequipped) site at Katale was too great for many of the refugees: hundreds died of dehydration during the long walk up over the lava fields towards Katale, whilst tens of thousands decided to settle at Kibumba, a wholly inappropriate location, 27 km from the nearest water source, Lake Kivu.

The sites that were initially established were therefore a mixture of planned camps (Katale, Mugunga), where one of the primary criteria for selection was availability of water, and a more 'spontaneous' site at Kibumba, although even at Mugunga, refugees settled partly because they were following elements of the defeated Rwandan army (FAR) who had been allowed by the Zaïrean authorities to move west rather than north from Goma. A major problem was caused across the whole of the Goma area by the nature of the terrain, which was predominantly recent volcanic lava flows: these made it difficult for vehicles to leave the road; for buildings to be erected for services such as health clinics and food and water distribution centres; or for waste to be disposed of. At Kibumba, the water supply situation remained critical, with the failure of attempts to drill for water leading to a continued need for 1.5 million litres of clean water per day to be supplied by tanker (Biswas and Quiroz, 1995).

Of the three principle refugee-receiving areas, only in Bukavu was the pattern of settlement of the refugees more dispersed. Here, initial influxes of Hutu refugees in July 1994 moved to three planned camps which had been established earlier in the same year at Nyakavogo, Nyangezi and Nyatende for Tutsis who had fled at the start of the genocide in Rwanda, and who now rapidly returned to Rwanda. Subsequently, from July to October, a number of additional sites were identified by the UNHCR, assisted by NGOs such as Care, Oxfam and MSF. New arrivals, who came at a slower pace than in Goma, were moved out to these sites from Bukavu town. Borton et al. (1996) note that the process of site selection took some time, because of inadequate preparedness, lack of personnel, lack of suitable sites and problems in securing permission to use the land from local authorities and land owners. None the less, sites were found, and within a relatively short period.

Commenting on the establishment of refugee camps in north and south Kivu, Biswas and Quiroz (1995: 73) note that: 'The sudden arrival of a large number of refugees required difficult but rapid decisions on the siting of refugee camps to accommodate them. In retrospect, some of the decisions left much to be desired, especially in terms of the environmental cost.' Looking at these three areas of refugee settlement as a whole, only one site stands out as having been placed in a completely inappropriate location – that of Kibumba – although the location of Katale on the border of the Virunga National Park, a World Heritage Site,[2] might also be questioned from a conservation point of view. However, many settlements did grow much larger than might have been planned, and in this context it is worth noting that the speed and size of the influx in each area was strongly related to the final pattern of settlement. Thus where the influx was large and rapid, as in Goma and Ngara, the *de facto* result was large and arguably less suitable sites for settlement, whilst in Bukavu, where the agencies and local authorities had more time to plan, smaller and, on the whole, more appropriate sites were eventually found.

Although this conclusion points to the unavoidability of large camps in certain circumstances, two provisos must be made. First, it is important to bear in mind the role of the UNHCR, NGOs and government authorities in channelling new refugees to existing camps, even when these are clearly 'full' and cannot cope with additional influxes. For example, Jaspars (1994) points out that in spite of the fact that the majority of refugees entered Ngara District at one point (Rusumo Falls) on one day (28 April), a significant number also came via a range of other border crossings in the weeks and months that followed, and that when these people were 'discovered', they were encouraged to move on to Benaco, despite the overcrowding there. Of course, this partly reflects the determination of the Tanzanian government to avoid the local settlement of refugees, for example, in Ngara town. Indeed, in Goma, the DHA (1995: 3), in its appeal for funds on behalf of the UN to the international community, puts a different gloss on the directing of refugees to the north of Goma at the start of the influx, saying that this was a decision of the government.[3] It is interesting to note that in the Karagwe District of Tanzania, host by October to 155 000 refugees, who arrived at a similar pace to those in Ngara, there was no attempt to congregate all the refugees at one site, partly perhaps because of the isolation of the area and the much smaller presence of international agencies, and partly because a higher local population density restricted the (physical and political) space available to establish large camps for Rwandans only.

In addition to the issue of how large camps become established, there is also the question of why refugees were not subsequently dispersed to smaller sites once the emergency phase had passed and there had been time to identify alternative sites. Across the region, there were calls from representatives of international agencies to promote such dispersal (*cf.* ERM, 1994; Adisa, 1996). In the case of Tanzania, some new sites were found in order to reduce population densities, although the first two – Lumasi and Msuhura – were both almost adjacent to the main Benaco camp, whilst Kitali Hill was rapidly filled with new arrivals from Burundi. Another planned site, to the south of Lumasi, was abandoned even after feeder roads had been constructed because it was found to be on a mining concession and the owner objected. There are reports of a further potential site in Biharamulo District being abandoned after a freelance consultant working for an international conservation group found elephant dung and argued that the site would obstruct a migration route for elephants (M. Owen, 1997, pers. comm.).

Government proposals to move refugees to new sites any further away were always dismissed, probably rightly, as unrealistic by international agencies operating on the ground. For example, in a visit to Ngara in November 1994, President Mwinyi of Tanzania proposed relocating the Rwandan refugees who did not wish to repatriate to sites in the south of the country recently vacated by Mozambican refugees, although such a drastic relocation was opposed by international agencies, the refugees themselves, and the residents of southern Tanzania, and was never likely to occur on grounds of cost alone. Similarly, the Zaïrean authorities offered to move the Rwandans to available land around 100 km from the border; a more reasonable proposition, although the cost of this too was seen as prohibitive (Anacleti, 1996). In practice, 'decongestion' was achieved by establishing new sites at Kahindo, close to Katale, and at Lac Vert, adjacent to the site of Mugunga camp.

The key element in this discussion is the fact that finding land for refugee settlement is increasingly difficult. As a result, in the short term, issues surrounding the appropriateness of sites may be ignored in the practical search for land that is not ruled out for other reasons. Thus Anacleti (1996: 305) notes that in the Great Lakes region, for example, both Tanzania and Uganda still have 'large tracts of land' officially 'gazetted' for use by refugees. However, even though population density, for example in Ngara District, remains low, individual citizens now treat land increasingly as a commodity. He suggests that as they have seen more and more of their common property resources given by governments to individuals and private firms without their consultation, this has hardened their attitudes towards accepting government allocation of further land, even for 'good causes' such as refugee settlement. In addition, it seems clear that once refugees were settled in Benaco, a certain institutional inertia set in, and it became increasingly difficult to move them elsewhere. This was arguably reinforced by the investment in physical infrastructure at supposedly 'temporary' refugee camps made by the international agencies, not to mention the dependence of such agencies on the 'visibility' of refugees in order to ensure continued funding of their operations (Harrell-Bond et al., 1992).

In north and south Kivu, a rather different situation existed, in the sense that the refugees moved into areas that were already amongst the most densely populated in the country. Here, competition for land was already fierce, and ongoing conflict over land rights between the 'native' Zaïrean Hunde ethnic group and the Banyarwanda group, whose own origins were in Rwanda, was inflamed by the arrival of the refugees, many of whom sided with Hutu Banyarwanda against the Hunde and Tutsi Banyarwanda. Coupled with the desire of the international community to limit movement of the refugees into the Virunga National Park (and other national parks), this clearly limited the amount of land available for settlement. In both Tanzania and Kivu, it is interesting to note the assumption that refugees 'cause trouble', which underlines the difficulty of finding appropriate sites; and the fact that where environmental arguments were developed, these increased rather than decreased the difficulty of site selection.

However, it is also interesting to note another side to the story: in Kivu, because of the high population density, and the highly fertile soils in much of the region, land also has a high potential monetary value. This in turn meant that some land *was* made available by landowners, particularly in the Bukavu area, in the hope of extracting suitable 'compensation' from the UNHCR for its use. In many cases (including sites at Muku, Mushweshwe, Nyantende, Izirangoabo, Chidodobo, Bideka and Idjwi Island, all in south Kivu), refugees were settled adjacent to local villages, whose inhabitants also benefited from the services (water, health, etc.) provided for the refugees (Zabala, 1994). In general though, the apparent preference of environmentalists for smaller, more dispersed camps, and such potential socio-economic benefits of these for local people, were not combined into a significant push for camp dispersal.

Initial environmental concerns: the conclusions of EIAs

Whatever the causes of the establishment of large camps for the majority of the Rwandan refugees in Tanzania and Kivu, the result was that they were established, and calls for them to be dispersed on environmental grounds were likely to fall on deaf ears. Given this situation, this section moves on to discuss the 'environmental impacts' identified by a wave of reports and environmental assessment teams, before considering how these reports were translated into environmental policies aimed at mitigating the identified impacts. As might have been expected from the review of existing concerns in other refugee-affected areas in Chapter 2, and past policies considered in Chapter 4, 'deforestation' ranked high amongst the environmental problems, although issues concerning water, waste disposal and possible impacts on threatened ecosystems or species have also received attention.

All of the contemporary reports written on the environmental consequences of the Rwanda tragedy highlighted 'deforestation' as a major consequence. For example, an initial EIA conducted for the UNHCR by Ketel (1994a: 10) cited 'depletion of forest reserves' as the only 'direct impact' of refugees, although a number of 'indirect impacts' were also listed. A subsequent report on north and south Kivu by the same author (Ketel, 1994b) qualified this statement, noting that in Ngara, Tanzania, the tree savannah ecosystem possesses a certain capacity for resilience and regeneration, but continued that this was not the case in Kivu, where in the Virunga and Kauza-Biega Parks, both affected by refugees, there was 'no possibility of regeneration' of the park vegetation to its original state, concluding that 'In both cases, Tanzania and Zaïre, the impact is direct, negative, and very visible' (Ketel, 1994b: 28). Similarly, Biswas and Quiroz (1995: 73) report on their mission to Kivu for the UNDP that 'The most serious environmental problem created by the refugees has been deforestation around all major camps.'

Given the emphatic nature of these statements on the severity of deforestation, surprisingly little hard evidence was presented by their authors of actual rates of loss of woodland or forest – whilst there was also little discussion of sustainable levels of harvesting, or of the relative importance of the different economic or ecological significance of different types of woodland (e.g. tree savanna, riverine forest, montane forest, etc.). For example, Biswas and Quiroz (1995) cited 'government estimates' that 3 758 ha of forest land was lost in south Kivu within three weeks of the arrival of the refugees, along with 19 ha in Bukavu city, and 300 ha in Goma. They went on to state that in the Virunga National Park, 7 200 ha had been deforested in the Rugo-Kibati sector of the park, and 4 675 ha 'devastated' in the Mugunga sector, 'primarily by heavily armed Rwandan Army personnel', apparently basing these estimates on a GTZ study of woodfuel collection over five days in September–October 1994. The UNHCR's 'environmental co-ordinator' in Goma provides a different figure, with 3300 ha of 'dense rainforest' apparently removed in the first year of the influx (based on analysis of satellite imagery), compared with an estimate of 1700 ha in the second year, presumably indicating the 'success' of the UNHCR's environmental projects (Leusch and Burie, 1996).[4] Ketel (1994a) gives no estimate of forest lost in Tanzania, although a subsequent project appraisal report for GTZ in the area concluded that there was an annual deforestation rate of 23 000 ha.

However, the usefulness of most of these estimates can be questioned. In Tanzania, estimates were based on an extrapolation of woodfuel consumption of the kind criticised in Chapter 2, where the assumption is that consumption in cubic metres of wood translates directly into deforestation, with no account taken of the use of dead wood or of the regeneration capacity of the ecosystem. In Goma, figures on deforestation provided by Leusch and Burie (1996) were based on a detailed analysis using satellite images for 1994 and 1995, but a closer examination of this analysis provides a rather different picture. Thus around Kahindo and Katale camps, only 1 ha of forest land inside the Virunga National Park had been 'totally deforested' by July 1995, with 52 ha thinned to 30 per cent of its original density, and 792 ha thinned to 75 per cent of its original density (UNHCR/GTZ, 1995). This was calculated to imply an 'equivalent area deforested' (*équivalent rasée*) of 235 ha, although this rises to 467 ha if the area outside the park is included. In Mugunga–Lac Vert, a slightly different calculation was made, based on the type of forest lost, showing the 'equivalent' 497 ha of 'dense' forest lost, and 1103 ha of 'open' forest. The point is that there is a need to differentiate estimates of 'deforestation' according to the density and structure of vegetation that was there before, and the extent to which it has been removed, whilst analysis might also be extended to consider what type of vegetation has replaced it. Thus in a number of cases, 'deforestation' is more correctly termed 'conversion of forest to agriculture' – an issue which is returned to below.

Without the benefit of satellite imagery, a more guarded analysis was also provided for Tanzania by a comprehensive EIA conducted for the agency Care by Environmental Resources Management (ERM, 1994). Whilst providing an estimate of annual deforestation of a similar order of magnitude to that later produced by GTZ, the report noted that only around 30 per cent of households in the larger camps, and few in the smaller camps, were burning 'green' wood at the time of the assessment. For a few months at least, the report concluded that stocks of dry wood are 'probably adequate to meet initial total demand', although it noted that this situation was likely to change over time (ERM, 1994: 33). In addition, the ERM team conducted a survey of woodfuel consumption which found very high rates of woodfuel consumption, at a average of 2.8 kg per person per day. This figure was confirmed by subsequent monitoring of consumption and wood brought into the camps by Care (J. Murray, 1997, pers. comm.; see also Owen, 1996a). However, this can be interpreted in two ways: either the refugees were using wood in a highly wasteful way, thus contributing to excessive deforestation, or there was simply no pressure on wood supply that would lead to more economic use. In practice, both conclusions were supported by the woodfuel survey of ERM, which found a significantly higher level of woodfuel consumption by refugees compared with local people in the northern district of Karagwe, where wood is more scarce, but no significant difference (at the 95 per cent confidence level) in the southern district of Ngara, where wood is more abundant. Once again, the small size and limited time period over which the survey was conducted should lead to caution in the interpretation of its results.

In addition to deforestation, a number of other negative environmental impacts were cited by the various EIAs conducted in the region, although again, the amount of evidence presented to back up claims of environmental degradation is limited. One of the few assessments of the Rwandan refugee emergency to attempt to put a cost on the environmental damage caused by refugees was that of Green (1995) in a study conducted for UNICEF. Although recognising that assessing the costs of environmental degradation is difficult, Green concentrated on costs in the form of the added workload on women and girls of travelling further to collect food and water. This he estimated at around 20 000 extra woman-years per annum (one and a half hours per household per day), or $4 875 000 – although these are costs are 'notional', and incurred by refugee women rather than by host populations (or by assistance agencies, which perhaps more clearly explains why this kind of calculation is unusual!).

Returning to impacts on natural resources, Biswas and Quiroz (1995) reported accelerated soil erosion – itself a consequence of deforestation – in and around Goma, and more especially around Bukavu, where steep slopes containing alluvial soils were exposed, leading to gully erosion. However, soil erosion is cited by Kunze et al. (1990) as a major problem prior to the arrival of the refugees in Bukavu, calling into question whether the observations of Biswas and Quiroz were able to distinguish additional erosion caused by the refugees' presence from a process that was already occurring. Another relevant observation is that of Shapiro (1989), who noted a shortening of fallow periods in various parts of the Kivu region prior to the arrival of the refugees, which had also already affected soil structures. The observation of reduced fallow is supported by Fairhead (1990) for north Kivu, although in this case an explicit link with the physical removal of soil is not made. Ketel (1994a) also cited soil erosion in Tanzania as a result of the refugees' presence, although ERM (1994) was more circumspect, noting again that soil erosion was already an ongoing process in the region, and that the rate of accelerated erosion in the future would depend on the type of land cover established in the medium term.

The ERM study placed more emphasis on degradation of water sources, an issue mentioned by Ketel (1994a), but rejected (1995: 75) for Kivu by Biswas and Quiroz, who concluded that 'Thus far, there is no specific evidence that the refugees have significantly contributed to a deterioration in water quality'. Thus the latter authors dismissed reports that the refugees were responsible for polluting Lake Kivu, identifying high concentrations of phenol as the major source of contamination, which was unlikely to be linked to the presence of the refugees. None the less, as noted above in the section on site selection, water quantity – or lack of water – continued to be a serious problem in both Benaco and Kibumba camps throughout their two and a half years of existence. In Benaco at least, ERM (1994) suggested that this might have had a long-term effect in terms of lowering the yield of springs and shallow boreholes, whilst this was coupled with a possible contamination of water sources where latrines were in direct contact with fractured shales.

Of all of the many environmental impacts noted though, deforestation took centre stage. Nowhere was this more contentious than in the Virunga National Park to the north of Goma. The park was initially established by the Belgian King Albert in 1925, as the *Parc Nationale Albert*, and was Africa's first national park. It is of particular conservation importance because of the population of around 300 mountain gorillas which live on the slopes of three volcanoes in the extreme south of the park – directly adjacent to the zone into which the Rwandan refugees moved in 1994. Quite apart from the risk to the refugees of eruption of the volcanoes (the last eruption of Nyamuragira was in 1984, and a Japanese vulcanologist predicted it would erupt again in 1995), progressive deforestation of the mountain slopes threatened the habitat of the mountain gorillas, which are already a globally endangered species. Alarm bells were raised soon after the refugees' arrival in the region by a rapid assessment by the Technical Unit of UNHCR Goma, which identified large-scale incursions into the park of up to 40 000 refugees each day, as well as commercialisation of products harvested there, especially wood, which was transported to nearby Goma town.

However, once again, considerable complexity to this situation is worth noting. First, the history of development of the park, whose boundaries were expanded in 1927, 1929 and 1935, has been characterised by tension between the state and conservationists on the one hand, and various local populations on the other. For example, Fairhead (1990: 89) describes how the Belgian authorities took advantage of the temporary relocation of populations in what is now the northern part of the park, who had been forced to move to higher altitudes by outbreaks of sleeping sickness, to declare the area 'uninhabited' and therefore suitable for inclusion in the park's boundaries. Although compensation was paid to local leaders for land taken by the park, it seems that little of this found its way to local populations. The complexity of claims on the land of the park was also reflected in its unusual division into 'integral zones', where settlement was not allowed, and 'annexed territories', where populations remained. This system had collapsed well before the arrival of the refugees, but none the less reflected at least an awareness by colonial authorities of the need to address both conservation and indigenous rights concerns.

Thus issues surrounding settlement have remained contentious between local populations (many of whom believe they were dispossessed of their land) and first the colonial authorities, followed by the Zaïrean government and international wildlife organisations, who have sought to protect the gorillas. In turn, these conflicting claims have been far from being regulated or under the control of the government or the rule of law, with Bobb (1988: 223) noting that: 'remote and dense, the Virungas have been a haven for anti-government guerrillas since the early 1960s. The inhabitants of the region are said to remain still largely outside the sphere of influence of the government.' Although this latter point probably referred to parts of the mountain range away from the immediate area of settlement of the refugees, the arrival of the Rwandans and their settlement next to the Virunga National Park represented in part an opportunity to 'reclaim' lands lost by local communities, not least due to ethnic ties between the Rwandan refugees and the local (Hutu) dispossessed.

At the same time, the whole notion of Kivu as a region that suddenly became overpopulated in 1994 (as implied in some reports on the refugee influx) is hardly tenable. Thus in their preparation of a Strategic Environmental Action Plan for the region for the UNEP and Habitat, Sheckler et al. (1996: 40) note that across the Great Lakes, 'major environmental changes were in place even before 1990', and although they focus on overpopulation as an issue, they also note that 'much of the forests of Kivu province, Zaïre were depleted early in this century for coffee and tea planting'. This theme is also stressed by Witte (1992), who identified expropriation of land by the state élite since colonial times, and its granting as concessions for ranching, as well as plantations of coffee, tea and cinchona, as major causes of deforestation. More importantly, Witte noted how in order to provide labour for these plantations, people were relocated into Kivu from Rwanda and other parts of Zaïre. As part of this process, land was forcibly reallocated to these immigrants. The land grant system which allowed this to happen was still in force (and in use) when the refugees arrived in 1994. Nor is this process restricted to Kivu: for example, the Virunga Mountains form a continuous ecological zone that spans across to neighbouring Rwanda. There, Waller (1993: 29) describes how the expansion of pyrethrum farming had already significantly reduced the mountain gorillas' habitat, as projects funded by the European Development Fund had led to the loss of 'half of the forests'. This is the region that many of the Rwandan refugees around Virunga had fled from – and they too had been affected by this loss of land to commercial (or as it turned out, not commercially viable) plantations.

Turning EIA into environmental policy

After the initial flurry of environmental assessments in the region surrounding Rwanda, the task remained of converting their conclusions into environmental policy, in the form of an environmental action plan (EAP). In Tanzania, an initial step was taken soon after the refugees' arrival with the establishment of an 'environmental task force' (ETF), which was charged with helping to draw up an EAP, under the guidance of a UNHCR environmental co-ordinator. The task force was set up by the UNHCR and sought to draw together all those with an interest in environmental matters, including international and local NGOs funding or implementing environmental projects, as well as representatives of the district and regional administrations. In turn, funding was rapidly made available for environmental activities, notably by IFAD, through a contract supervised by the UNHCR environmental co-ordinator and implemented by Care. In Goma, an environmental strategy appears to have been put forward (in September 1994) by the local UNHCR technical unit responsible for all site planning activities, even before an EIA had been conducted (in November 1994). It was only after this that an environmental co-ordinator was appointed within the technical unit, and an ETF established. Funding for environmental activities, which had been initiated at the start of the influx by GTZ using funds from the German government (BMZ), was subsequently included in the 1995 UN Consolidated Appeal for the region.

In one sense, the task forces established in Kagera and north Kivu could be criticised as simply being 'talking shops', with no real power to influence wider policies that had important environmental consequences. For example, in Tanzania at least, the UNHCR and government personnel responsible for site identification and planning were not part of the task force; at the same time, in neither country was there formal representation of refugees on the ETF, although in Tanzania, refugee environmental committees were established subsequently within the camps themselves. Implementation of environmental activities meanwhile were the responsibility of implementing agencies, which were not required to listen to the recommendations of the ETF; indeed in practice they found themselves highly limited by the budget categories allocated at the time of funding, such that they would have been unable to respond to significant changes of approach even if these had been recommended by the ETFs. However, the Rwandan emergency represented the first (and to date only) refugee emergency in which environmental co-ordinators and task forces have been deployed, making their experience particularly interesting in terms of potential development of environmental policy in emergency situations.

In the case of Tanzania, the ETF initially pushed forward with the task of supporting the development of an EAP, providing a forum for debate between the principal agencies involved in environmental activities (Care and GTZ) and at least some other concerned parties. Since the ETF was established prior to the ERM study, the ERM consultants were able to benefit from participating in task force meetings in developing their proposals for environmental mitigation measures – although other studies were less connected to the ETF. In turn, the overwhelming concern of initial EIAs with deforestation was reflected in a draft EAP which was quickly drawn up in Tanzania, in which eight out of nine areas of action were directly related to deforestation (Table 5.3). Of these, by far the most important was a massive operation to supply woodfuel to the camps described below. In addition, a relatively standard approach included provision of fuel-saving stoves, a search for alternative energy sources and sundry environmental rehabilitation activities. Interestingly, though, the initial plan drawn up by ERM did not include significant provision for the establishment of tree nurseries or for large-scale reforestation activities: indeed, ERM (1994: 50) commented that reforestation was 'almost certain to fail on most soils in the vicinity of the camps'. Perhaps more worryingly, this initial plan also had little to say about protected areas that were located close to the refugee settlements, despite the fact that the Msuhura camp was just a few kilometres from the border of the Burigi Game Reserve, and poaching of game in the reserve was already starting to be an issue of concern at the time of the mission – at least to agency workers who spent days off travelling around the area to view animals![5]

In contrast, in north Kivu, as noted above, encroachment into the Virunga National Park was one of the principal concerns in the area, with the result that the evolving environmental strategy concentrated much more on issues of protection and enforcement of environmental regulations than in Tanzania. This included the establishment of an Environmental Information Bureau in Goma by the UNHCR and the German agency GTZ, to monitor threats to the national park, which aimed to raise awareness of environmental issues both internationally and amongst agen-

Table 5.3 Draft UNHCR/NGO Environmental Action Plan, Tanzania, 1994

Activity	Actions to be taken
1. Tree marking	Living trees and shrubs of desirable properties marked to prevent cutting
2. Identification of wood sources	Forest areas which are well stocked with dry wood, and sites suitable for cutting of green wood identified
3. Harvesting and transport of wood	Promote regular supply of wood to refugee camps. Wood cut under supervision in selected sites, leaving sufficient density for regeneration. Wood transported to sites around camps
4. Protection of natural water sources	Identification, rehabilitation and protection of natural water sources (e.g. with tree planting)
5. Building wood	Small-sized wood, reeds and papyrus harvested under controlled conditions and provided for hut construction
6. Improved woodfuel stoves	Wood-saving stoves constructed using simple locally available materials
7. Alternative energy sources	Identification and testing of energy sources other than woodfuel (e.g. peat, charcoal briquettes)
8. Rehabilitation	Replanting of seedlings, cuttings and direct sowing of seeds to reforest in areas identified by local populations
9. Environmental education	Promotion of environmental awareness in refugee primary and secondary schools

Source: Environmental Task Force, Ngara, October 1994.

cies working in the region. Assistance was also provided to support the park authorities and the Institut Zaïrois de la Conservation de la Nature (IZCN) by the UNHCR, the European Union, the UNDP, UNESCO, the World Wide Fund for Nature (WWF) and other donors, including activities to control wood cutting and poaching. In addition to these park-related interventions, at least five different projects sought to disseminate fuel-saving stoves; a comprehensive survey of forest plantations was conducted; a waste disposal system was installed; tree nurseries were set up, funded by the UNHCR, GTZ and WWF; and most important of all, a wood-fuel supply project was established to provide wood to the refugee camps.

In analysing the record of the environmental co-ordinators and task forces in these two regions, it is important to start by noting that the initiatives of the ETFs in Kagera and north Kivu were not the only sources of environmental policy, or of policy based on environmental reasoning. First, although the ETFs played an important role in bringing together different agencies, they had no formal power to regulate the activities of other agencies. As a result, these agencies could (and often did) act independently if they wished. Such action had both positive and negative results. For example, in both Goma and Kagera, environmental activities had already been initiated by one agency, GTZ, prior to any assessment missions being sent to the region, or co-ordination structures being established. Thus in Kagera, a household energy programme that had been operating inside Rwanda continued its activities in the camps with staff who crossed the border as refugees, and a budget that was transferred directly from Rwanda to Tanzania. It was this project which initiated tree marking to conserve larger trees around the camps – an activity which, although included in the ETF, had already been more or less completed by the time the latter was drawn up.[6] It was this project too that had already disseminated thousands of mud-stoves to reduce fuel consumption by the time of the ERM study, although as was noted in Chapter 4, these stoves did not appear to have led to any significant reduction in wood consumption by the families who used them.[7]

In other cases, though, there was duplication of activity, or worse, parallel policy decisions which contradicted the sense of what the ETF and environmental co-ordinators were trying to achieve. For example, it was perhaps ironic that whilst environmental specialists were trying to reduce refugee dependence on local supplies of woodfuel, the DHA (1995: 24) reported in the second inter-agency appeal for the Rwanda emergency that for shelter, 'imported items (i.e. plastic sheeting) will be replaced by local materials where possible' – presumably implying increased reliance on precisely the same wood supplies for which the ETF was seeking to reduce demand. The UNHCR's environmental specialists in Tanzania and Kivu also appeared to have scant influence on broader policy decisions by the UNHCR or other international agencies, which probably had a much more significant effect on 'environmental impacts' – such as on the composition of the food basket, the type of cooking utensils distributed, or the planning of new contingency sites.

A further element of independent action was that of the host governments. For example, since formal power rested with the government, the government increasingly sought to use it, although in both cases its capacity to act independently of the international agencies was severely limited. Thus in Tanzania, it was the central government, not the ETF, which introduced a rule in 1996 that refugees were not allowed to move further than 4 km outside the camps in Ngara District, and even proposed that the UNHCR should physically demarcate the camp boundaries to indicate this restriction. The proposal was opposed locally by several international agencies, and was never convincingly implemented. Similarly, the Zaïrean government announced 'closure' of Kibumba and Nyangezi-Mulwa camps in February 1996, implying a 'ban on all commercial and non-essential activity inside the camps, which had become small cities with shops, restaurants, markets and schools. In addition, refugees were prohibited from moving in and out of the camp, and thus from seeking employment outside. Zaïrean troops surrounded the camp whilst others went inside to enforce the closure of schools and "businesses"' (EIU, 1996a: 25). In this case, environmental and security initiatives intertwined to produce policy, although once again, enforcement was limited.

None the less, despite the task forces' limited powers and room for manoeuvre in changing the orientation of policy that had already been set, either at the start of the emergency or at the time of project formulation, they do appear to have played a useful role in certain respects. Perhaps most importantly, they provided a forum for discussion of policy by agencies that are often excluded from any role. This was particularly noticeable in Tanzania, where local government had otherwise been largely ignored by the (almost exclusively international) aid agencies that were operating in the region, and had little capacity to participate in project design or implementation as a result of critical shortages of funding and transport. For the first time, government officials in the district and regional administrations were able to participate in at least a form of decision making, whilst the task force (or perhaps more cynically, the large budget allocated to environmental activities) also provided a stimulus to local NGOs, which had also not featured significantly in the first few months of response to the crisis.

Woodfuel supply

Although the environmental strategy proposed for both Tanzania and Kivu during the Rwandan emergency had various elements, perhaps the most important in both countries was the initiation of large-scale projects to supply fuel to the refugees. Thus a major part of the environmental project funded by the Belgian Survival Fund through IFAD, and implemented by Care through the UNHCR in Ngara District, Tanzania, involved the supply of woodfuel to camps (Figure 5.4). In Goma, attempts were also made to supply refugees through a woodfuel provision programme initiated by GTZ from the start of the emergency, and subsequently implemented by the UNHCR in collaboration with Care and GTZ (Umlas, 1996b). In Tanzania, the intention was to cover almost all the refugees' consumption needs (Grimsich, 1996) through controlled cutting and provision of wood (and some other fuels, notably peat). In north Kivu, the aim was to supply as much wood as possible, recognising that this was only likely to meet about 20–30 per cent of refugees' needs.[8]

These programmes were increasingly being called into question by the time of the mass return of refugees to Rwanda in late 1996, reflecting their high cost, and a number of management difficulties. For example, the supply of woodfuel to the camps in Tanzania had consumed some $2.5 million by late 1995 (Grimsich, 1996), whilst Owen (1996b) estimated that $8.6 million per year would have been required to supply all the refugees with just 1 kg of wood per day.[9] In Goma, woodfuel supply was estimated at one stage to be consuming some 20 per cent of the entire aid budget to the region. Both projects were based on the assumption that by providing sufficient wood, pressure would be taken off woodlands adjacent to the camp, so that tree cover could then be maintained at a density sufficient to allow natural regeneration. The woodfuel that was supplied

Figure 5.4 Controlled harvesting of wood, Ngara District, Tanzania, 1994
Source: (photo: Richard Black)

was harvested in a 'sustainable' manner from sites at some distance from the camps, and brought to drop-off points outside the camps by truck. Despite these good intentions though, there was little evidence that this strategy had a significant effect in reducing pressure on woodlands in the area. In Goma at least, there appears to have been an inequitable pattern of distribution so that some households received no wood whilst many received insufficient quantities; there were also reports of widespread corruption amongst truck drivers, as well as fatal accidents resulting from their disrespect for the camp speed limits (Kimani, 1994). One consultant firm working for GTZ pointed out that the 'sustainable harvesting' of wood from private wood farms, in addition to being expensive, also increased levels of consumption since much of the wood supplied was 'green' (HEAT GmbH, 1994).

It is possible to draw some meaningful conclusions about the effects of the woodfuel supply project in Tanzania, thanks to detailed studies of energy consumption and demand conducted by two UNHCR consultants, Matthew Owen and Ivan Ruzicka, as well as earlier monitoring of woodfuel use by Care, the implementing agency for the project. These studies showed that woodfuel consumption did indeed decline over time in the Kagera camps (and over the duration of the project) from over 2.5 kg per person per day at the start of the emergency to around 1.5 kg per person per day by October 1996. However, consumption, which varied widely between different camps, was found to correlate strongly with the price of woodfuel and other market indicators, and in particular, when the programme ended in late 1995, there was a significant fall in consumption (Owen, 1996a). Essentially, in camps where woodfuel was scarce, or where there was strict enforcement by government authorities of rules prohibiting the cutting of trees in the vicinity of camps, the market price for woodfuel was high and consumption low. However, where 'sufficient' wood was supplied by the UNHCR/Care project, there was no incentive to reduce consumption, and consumption remained high.

Environmental outcomes

The Rwandan refugee emergency lasted a total of two and a half years in Tanzania, and slightly shorter in neighbouring Kivu – although in the latter case residual populations of refugees and internally displaced people still exist, whilst in the former, attention has shifted to the situation of some 400 000 refugees in the Kigoma region to the south, with many recent arrivals from Burundi and the new Democratic Republic of the Congo (DRC). Such a time-span, though long for those involved, was relatively short by African (and indeed world) standards; after all, the previous group of Rwandans to seek refuge in Tanzania came between 1959 and 1964, and were mostly there until the new exodus of 1994. In turn, given such a relatively short stay, it might be argued that the worst predictions of environmental catastrophe did not occur because the pressure of this increased population on the environment and natural resources was released at a critical time. After the refugees' return at the end of 1996, it was possible, in Tanzania at least, for the environment to start to regenerate to its former state, even though this might prove a long process.

However, despite the rapid and somewhat dramatic ending of the Rwandan refugee emergency in neighbouring countries (internal displacement inside Rwanda remains a big issue), it is still interesting to look back to see whether some of the more dire predictions of environmentalists at the time of the emergency proved true – or would have been likely to had the refugees' presence continued. Looking first at Tanzania, an initial point is that it is quite clear that a large area that was previously *Brachystegia* and *Acacia* woodland was cleared. For example, Grimsich (1996: 50) states that by 1996, an area of 12 000 ha was being cultivated by refugees, most of which would formerly have been woodland. None the less, whether this is considered 'environmental degradation' is a moot point, since much of the area cleared was subsequently put to agricultural use, either directly by the refugees, or by local people, using the refugees effectively as a (cheap) labour force. The development of agriculture, with the acquiescence or active encouragement of sections of the local population, also helps to put into perspective early statements about the destruction of crops by refugees in search of food on arrival, as well as other initial environmental pressures placed on water resources and physical infrastructure. Thus NRI/ODA (1995: 13) note: 'From the very beginning, the camps ... had a close relationship with neighbouring villages, and these relationships have not by any means, always been deleterious. Markets have been established and refugees have provided labour to adjacent farms.' Of course, one aspect of these developments may have been a worsening of the situation for poorer local hosts, some of whom were reliant on opportunities for seasonal employment on local farms. The extent to which wages and employment conditions were under-cut in 1994–96 in Tanzania is unclear, but Chambers (1986) describes how 20 years earlier, during the previous Rwandan influx, this was precisely the outcome. Another aspect is that access to fresh food had the potential to play a role in reducing household demand for woodfuel.

Beyond the issue of deforestation, the broader long-term environ-mental impact of the refugees' presence in Tanzania is much more difficult to gauge. On the one hand, a comprehensive judgement about long-term impacts must await evidence of whether the local ecosystem has indeed been resilient to the pressures placed upon it, although evidence suggests that natural regeneration will now occur. One process that might arrest such regeneration would be if local populations (and possible internal migrants to the area) maintain efforts to cultivate crops, switching from the refugee market to markets inside Rwanda (the nearest substantial concentration of population), although this seems highly unlikely. It will also be interesting to see whether any of the environmental interventions aimed at the longer term – notably efforts at reforestation – result in the replacement of indigenous woodlands with commercially viable forest plantations. On the other hand, despite early complaints of the damage caused to roads and other infrastructure by increased traffic of aid vehicles and physical use by refugees, it seems fairly clear that the road system at least that was bequeathed after the refugees' departure was considerably better than that which existed before, as a result of investments made during the refugee influx. Once again, the major question is whether the standard of infrastructure can be maintained now that the Kagera region is once again relatively depopulated.

In the Democratic Republic of the Congo, there is a further complica-tion in providing a post-return assessment of the environmental con-sequences of refugee settlement, in that the regions of north and south Kivu remain somewhat in turmoil despite the refugees' departure. At one level, it appears that despite the warnings – or alternatively precisely because the warnings were heeded – there was no accelerated decline in the population of mountain gorillas in the Virunga National Park, although whether pressure on this area has been released sufficiently to allow regeneration of the natural vegetation of the park, and an enlarge-ment of their habitat, remains to be seen. Meanwhile, in discussing envir-onmental outcomes, it is interesting to focus not only on the consequences of the presence of Rwandan refugees in neighbouring states but also to consider the consequences (and cost-effectiveness) of the policy interven-tions that were implemented in the region to try to avoid or mitigate negative environmental outcomes. The huge cost of woodfuel supply has been identified above; and yet the one comprehensive study of its impact suggests that 'natural' processes of market adjustment to scarcity were much more effective at regulating natural resource use (Owen, 1996b). Similar questions might also be asked of the cost of marking trees – which were respected when wood was abundant, but appear to have been ignored when it became scarce; or of promoting fuel-saving stoves, which were used by 70 per cent of the refugee population, but which appeared to have no effect at all on woodfuel consumption.

The lessons of Rwanda for environmental policy in emergencies

A number of tentative conclusions can be drawn from the Rwandan emergency, both in relation to the development of environmental policy in emergency situations and in relation to the wider literature on environment and sustainable development within which this book is situated. On the latter point, in Chapter 1, the extent to which environmental concerns are central to both development and humanitarian activities, or whether they are simply a convenient rallying point to climb onto the funding bandwagon, can be called into question. The Rwandan emergency provides evidence for both positions: on the one hand, considerable sums of money were made available for environmental activities, seemingly before the numerous environmental assessments had been carried out. The extent to which a genuine environmental problem existed in Tanzania, certainly outside the immediate vicinity of the large camp at Benaco, is debatable. This is particularly so if one views agricultural land as of similar or greater 'value' to the ecosystem it replaced; or if one considers the likely resilience of that ecosystem once the pressure of the refugees' presence is removed – as it was in December 1996. However, on the other hand, as time progressed, availability of woodfuel, and more especially water, did start to pose problems for refugees living in Benaco, whilst in north Kivu, the difficulty of access to basic natural resources was a major issue for the livelihood status of refugees in certain camps, although de Waal (1997: 206–7) notes that they remained 'very well cared for by African refugee standards', with 'excellent water and sanitation systems'. Indeed, Stockton (1996: 17) describes the paradox of Goma as one in which, whilst the failings of the international assistance regime in general have been highlighted, none the less, refugees received levels of provision that 'greatly outstripped the quality of public services available for the host Zaïrean population'.

From a more directly policy-orientated point of view, one point raised by Ketel (1994b) in his environmental assessment of the refugee influx to north and south Kivu, and reinforced in the discussion on site selection above, is that by the time any environmental assessment is conducted, all the major decisions regarding site identification have already been made, and site planning and development has mostly taken place or has been finalised. Ketel concludes from this, and the fact that *de facto* decisions had also been made in Goma about the system of fuel use at a household level, the nature of humanitarian assistance, and sources of building material, that there is a need for EIAs to be conducted earlier in the emergency phase. However, it seems unlikely that, however early an EIA is conducted, this would be likely to have any serious impact on either site selection or the nature of household energy, building material or humanitarian assistance decisions. It is more a question that principles should be established about types of sites to be avoided, or patterns of assistance or organisation that work best for the environment (and the subsistence needs of the population). The fact that the Kibumba site in north Kivu was established and was not relocated is testimony to the difficulty of following even such a simple guideline of ensuring availability of water.

Another key policy point concerns the efficacy of woodfuel supply in refugee situations – an issue initially raised in Chapter 4. The evidence from Tanzania suggests that such a strategy is fatally flawed since it is likely to have the effect of reducing pressure to cut wood consumption. None the less, the Great Lakes emergency as a whole provides mixed signals on the subject, and suggests that in certain very well-defined circumstances, wood supply may be appropriate, or even necessary if nutritional standards are to be maintained. For example, the development of the Tanzanian woodfuel supply project was heavily influenced by apparently successful projects in Rwanda, which had already been implemented prior to the genocide. What is interesting about these projects is that they were concentrated in areas where the local supply of wood was highly limited, and refugees were forced in any case to buy wood on the open market. In this context, and unlike Tanzania, local demand for wood was largely irrelevant in determining local rates of deforestation, since the small areas of existing forest were tightly controlled by private owners. Meanwhile, those traders who might respond to market signals within the camp woodfuel market were in any case benefiting from inclusion in the woodfuel supply programme – with the probable exception of small-scale wood retailers at the camp level (most of whom were women). A similar situation seems to have occurred in Kibumba camp in north Kivu, where the woodfuel supply project was most effective both in supplying wood and in reducing rates of deforestation. Of all of the camps in north Kivu, Kibumba had the least independent access to wood and other natural resources.

The Rwanda emergency also raises the question of how far those dealing with a refugee emergency – whether from an environmental or from any other point of view – can or should be aware of the complex history of a region before developing policy interventions. For example, the dominant image of Mobutu's Zaïre in the Western media at least was of a poverty-stricken area (albeit as a result of government corruption), stretched to breaking point, both environmentally and socially, by the sudden arrival of starving Rwandan refugees. Seen as 'foreign' and therefore alien to the area, there was widespread agreement that the only appropriate solution was for these refugees to repatriate as soon as possible. Yet a rather different impression is provided by the Economist Intelligence Unit (EIU, 1996b: 22), which notes: 'Kivu, and the Masisi and Rutshuru regions in particular, has always been amongst Zaïre's most fertile lands and used to be a prime producer of cattle, beans, vegetables and dairy products. Estimated livestock in the region has fallen from 300 000 head in early 1993 to less than 100 000 in 1996.' Of course, this need for contextual information and an understanding of a region's history and politics is not limited to the field of environment, as is cogently argued for the Rwandan emergency by Pottier (1996). What makes the lack of historical and contextual knowledge of refugee assistance agencies more surprising in the case of the Rwandan emergency, though, was the fact that the area's history has been so intimately tied up with forced displacements, and indeed the actions of the UNHCR itself (Betts, 1969, 1984; Pitterman, 1984).

One recommendation of successive reports in Tanzania (*cf.* Ketel, 1994a; ERM, 1994) was that an appropriate response to environmental problems caused by the concentration of refugees in Ngara District would be to disperse them to other parts of Tanzania. To a certain extent, the history of previous Rwandan influxes into the country in the 1960s, in which they were largely dispersed and indeed offered Tanzanian citizenship, provided a model and justification for such a process. For example, the relatively favourable situation of Rwandan refugees in Tanzania during the 1980s and early 1990s can be contrasted with that in neighbouring Uganda, where Rwandan Tutsi refugees were neither dispersed nor integrated into Ugandan society (although many were integrated into, and trained by, the Ugandan army). It was from Uganda that, ultimately, Rwandan refugees launched the attack on Rwanda itself which installed the Rwandan Patriotic Front in power. In this context, it is interesting to note the call of Foster (1989) that such a dispersal and granting of citizenship to Rwandan refugees should occur in Uganda on the grounds of the environmental degradation being caused by their presence along the border. The case of Rwandans living in Uganda demonstrates the clear interconnections between environmental, political and security criteria in terms of refugee policy. It is also perhaps ironic to note that one of the principal arguments of the Tanzanian government in 1995 *against* refugee dispersal was precisely concerns over security in the region. Interestingly, this policy was partially reversed in the subsequent influx of Burundian refugees to the Kigoma region, when it was decided to locate camps at least 20 km from the border.

Another irony of the Rwanda emergency is that it should have been Tanzania – of whom Rutinwa (1996: 295) correctly writes: 'over the last thirty years, Tanzania's record of welcoming and accommodating refugees from neighbouring countries has been second to none' – that was by mid-1995 justifying the adoption of a 'no more refugees' policy, and closure of the borders with Rwanda and Burundi, with arguments that at least in part were environmentally grounded. Thus although security fears over the spread of the Rwanda conflict and the use of refugee camps as a base by the *Interahamwe* were real, and possibly more significant in guiding Tanzanian government policy (along with an increasing friendliness towards the new Rwandan and the Burundian governments), it was at least plausible to speak of population pressure (in one of the least populated parts of Tanzania), environmental damage (even though it was far from clear that this was irreversible), and damage to infrastructure (even though new roads and bridges had been built as part of the refugee assistance operation). Moreover, these allegations were, in Rutinwa's words (1996: 296), 'supported by several independent studies'.

   Finally, a greater focus on the historical perspective of the region might have also highlighted the significance of environmental change prior to the refugees' arrival in Kivu, as suggested in this chapter. Thus both north and south Kivu represented areas in which 'natural' vegetation had already been significantly disturbed prior to the refugees' arrival, to make way for plantation agriculture as well as the relatively high population density characteristic of the region. This was perhaps less significant in Tanzania, where NRI/ODA (1995: i) note that 'the pre-refugee situation in Kagera Region was one of relative stability'. However, even here, NRI/ODA pinpointed concern about the necessity for soil conservation that dated back to the 1930s and 1940s, such that 'the impact of the refugee presence may . . . be considered only to be one of degree and not significantly different from existing general problems associated with a growing population in a marginal area' (NRI/ODA, 1995: 14; see also Rounce, 1949). Moreover, the area had not been immune to significant population movement (the village closest to the refugee site at Benaco having been an *'ujamaa'* village to which people were relocated by the government in the 1970s), or to dramatic environmental change, with significant areas of woodland in the area having been clear-felled for control of tsetse fly at around the same time. Perhaps the most important factor in considering the 'environmental' impact of refugees on the Kagera region though was not its environmental vulnerability, but its remoteness in Tanzania. Indeed, it is important to place any environmental degradation that may have occurred during the refugees' presence in its wider context: an area that had previously been largely overlooked by international aid was by 1997 a recipient of large-scale intervention by the EU, the UNDP, IFAD, USAID, the UNHCR, the German agency BMZ, and the Japanese International Co-operation Agency (JICA). All of these agencies were effectively brought to Kagera by the refugees; all except perhaps the UNHCR are likely to stay, at least in the short term, despite the precipitous return of most or all of the Rwandan refugees late in 1996.

1. The film 'Exodus', in which the present author participated, was shown as part of BBC's 'Horizon' series in March 1995.

2. Since 1979, the Virunga National Park has had the highest legal protection status under the UNESCO World Heritage Convention.

3. In practice, in both Tanzania and Zaïre, attitudes towards the location of refugee settlements appear to have wavered. Both had histories of hosting dispersed populations of refugees, and both at various times proposed relocation of the refugees, albeit at prohibitive cost. To a certain extent, it could be argued that some individuals within the UNHCR and other agencies used government intransigence over site location as an excuse – citing the government's presumed position as the limiting factor rather than applying pressure to change the government's mind.

4. Unpublished figures from the UNHCR suggest that the level of deforestation did fall dramatically in Kibumba camp in its second year, and slightly in Katale and Kahindo, although there was a small increase in Mugunga and Lac Vert.

5. Grimsich (1996) suggests that up to 9 t of game meat were being removed illegally from the Burigi Game Reserve each week in early 1996, although the reserve did not contain any

endangered species or habitats. The presence of the refugees did affect revenues to the District Council from hunting concession fees and licences, which may have reached over $50 000 per annum prior to the refugees' arrival (M. Owen, 1997, pers. comm.). Also, as part of a corridor of protected areas stretching from the Akagera National Park in Rwanda through Kimisi, Burigi and Biharamulo Game Reserves to the Moyowosi/Kigosi complex to the south, the reserve did have a potential conservation interest, although Akagera is now largely destroyed.

6. It is arguable that the refugees would have left the larger trees which were marked by GTZ anyway, at least initially, since these were much more difficult to cut.

7. Studies conducted in Goma suggest in contrast that significant savings were derived from the dissemination of fuel-saving stoves, although these studies were almost exclusively conducted by or for one of the main implementing agencies, and so their impartiality might be questioned.

8. Defining the 'needs' of refugees for woodfuel is itself highly problematic, since actual consumption is likely to vary sharply according to supply.

9. This represents a cost of $14 per refugee per year, i.e. double the cost of kerosene provision to refugees in Nepal quoted in Chapter 4.

CHAPTER 6

Seven years on: refugees and natural resource management in West Africa

Although it is tempting to concentrate on short-term impacts of refugees on the natural environment, and policy responses that can be incorporated into emergency assistance programmes, it is arguably only in the medium term that the full extent of environmental change in refugee-affected areas can be assessed. This chapter therefore seeks to build on the conclusions of Chapter 5 by presenting the findings of recent research in two such medium-term refugee situations in West Africa, the Senegal River valley and the Forest Region of Guinea. These case studies exemplify the rather different concerns for environmental protection in the semi-arid zone of the Sahel and the West African tropical rain forest, respectively, but both cases stress the value of developing a flexible medium-term perspective towards environmental change. Freed from the sense of urgency and crisis inherent in the Rwandan situation, it is perhaps more feasible to examine competing interests in natural resources (for example, from different social groups and different genders), and the ways in which these might be acknowledged and accommodated by policy. Meanwhile, although the Rwanda crisis received – and continues to receive – considerable media attention, one of the striking aspects of the list of major refugee-affected countries provided in Table 2.1 is the remarkable lack of attention that has been paid to several other countries at the top of that list, where refugees have been present for a number of years. Nowhere is this more so than in the case of the Republic of Guinea in West Africa, host since 1990 to over 600 000 refugees, first from the war in Liberia, and then from the ongoing conflict in Sierra Leone.

Since the return of Rwandans in late 1996, the exodus to the Forest Region of Guinea now represents the largest single refugee flow in Africa – and the largest in the world to a single country to have occurred since 1990. Presidential elections in Liberia in June 1997 appear to have allowed the return of some Liberians in the dry season at the end of 1997, but prospects for return to Sierra Leone continue to look bleak. Indeed, at the time of writing, Sierra Leone continued to generate further refugee movements. It is in this context that this chapter sets out to examine the environmental consequences of the movement of Liberians and Sierra Leoneans to the Forest Region of Guinea, as well as considering the comparative case study of a much smaller (though similarly 'forgotten') flow of Mauritanian refugees to Guinea's neighbour, Senegal. A focus on these two case studies is of interest for a number of reasons. First, both these refugee flows occurred in the late 1980s or early 1990s, producing refugee populations that have existed long enough for a number of environmental (and other) impacts to be observable, but not so long that the populations concerned could be expected to have 'integrated' fully into their new host societies. Second, despite divergent histories since independence, both countries share a similar colonial heritage in terms of the institutional development of land and environmental policy, which has had ramifications to the present day. This is a heritage that is perhaps less familiar to many English-language readers, but of considerable importance to the development of conservation and 'sustainable development' policy across Africa as a whole.

In addition, the two case studies provide an interesting contrast in terms of the nature of the 'environment' into which these refugees moved: in the case of Guinea, refugees moved into a zone containing what are seen as the last remnants of West Africa's tropical rain forest; whilst in Senegal, the refugees were settled in the heart of the Sahel zone, in an area still recovering from the drought of the mid-1980s and in what is undoubtedly a highly vulnerable environment. Thus despite their contrasting environments, both case studies touch on issues of much wider public environmental concern, which have been specifically addressed in the post-Rio initiatives on biodiversity and desertification. This chapter is based on extensive fieldwork carried out in 1995 in the Yomou préfecture in Guinea, host to around 81 000 mainly Liberian refugees that year (Black and Sessay, 1995), and the Podor and Matam départements of Senegal, which between them took in over 47 000 refugees from neighbouring Mauritania in 1989, as well as a large number of Senegalese 'returnees' (Black et al., 1996). In both countries, interviews were conducted with a sample of both refugees and local populations, concentrating on three villages and adjacent refugee sites in each case. In both countries, the pattern of refugee settlement was 'dispersed', with refugee leaders negotiating settlement sites with village leaders at the time of the influx. Interviews were also conducted with these leaders, as well as with representatives of assistance agencies, and government officials at local, regional and national levels.

The Senegal River valley

Eight years after their initial arrival in 1989, Senegal was still host in 1997 to around 64 000 Mauritanian refugees (UNHCR, 1997), spread across over 250 'sites' along the Senegal River valley. Although now out of sight of much of the world's media, at the time the Mauritanian influx led to a major international assistance effort, involving a range of governmental and non-governmental agencies. The main element of international assistance to the refugees was the supply of food rations, including grain (usually sorghum or millet), beans, sugar, salt, cooking oil and powdered milk. From 1992 onwards, the size of the rations was progressively reduced, with the quantity of grain supplied declining by half between 1991 and 1993, from 500 g per person per day to 250 g per person per day. In addition to reflecting an expectation that the refugees would soon return to Mauritania, and pressure from donor countries to cut the cost of international assistance, this withdrawal of food rations was also largely based on the assumption that refugees had found land to farm locally and other income-generating activities, through which they would be able to achieve at least partial self-sufficiency (Thiadens, 1992). In this sense, and without pre-empting the discussion below, it could be argued on this basis alone that the 'environmental impact' of the refugees had been sufficiently contained by 1992. Thus the UNHCR was confident that sustained agricultural production could be achieved by refugees in the region, justifying the withdrawal of assistance.

Before moving on to questions of environmental impact, it is important to place the refugee influx into Senegal in 1989 in both its historical and environmental context. First, the refugee flow itself occurred against a background of deteriorating relations between the two countries, which led to an expulsion not only of refugees and Senegalese nationals from Mauritania but also of Mauritanians from several cities in Senegal. Considered by some to be a classic case of a conflict caused by competition over a deteriorating resource base (Kharoufi, 1994), disputes in the Senegal River valley, which divides the two countries, have a complex heritage. Initial clashes had their antecedents in tensions and conflicts spreading back over decades, notably between pastoralists and farmers; indeed, it was such a clash that sparked the events of 1989 which led to forced migration. In addition, Kharoufi has argued that discrimination against the black population of Mauritania can be traced in particular to a law in 1984 which abolished customary rights to land in the valley region. However, this law related not to competition over declining resources but at least in part to a significant *increase* in the value and productivity of land in the floodplain of the valley. This apparent paradox is explained by the construction of two dams (at Diama in Senegal, and Manantali in Mali) to regulate the flow of the river, and subsequent funding of a number of large-scale irrigation schemes between the two dams (Horowitz, 1991).

The importance of historical context can also be seen in terms of the pattern of migration in the valley region. The displacements of 1989 were clearly not exceptional, but rather can be seen as being superimposed on previous migratory patterns (Sane, 1993). Thus large-scale out-migration had become established as early as the 1940s, initially to Dakar and the Cap Vert peninsula, and later abroad, to a range of countries in Africa and Europe. Since the 1940s, consistently high rates of migration have been recorded in the region. Diop (1965) found that 70 000 Haalpulaar had migrated from the middle and upper valley in a study in 1959–60, whilst other demographic surveys have shown at least 20 per cent of the population of the valley 'permanently absent' (INSEE-Coopération, 1962; Dbaké, 1980), which excludes those who migrate on a seasonal basis. For example, a detailed study of a village near Podor (Lericollais and Vernière, 1975) found 344 emigrants absent, out of a village population of 1282 (i.e. 27 per cent).

High emigration reflects in part a shortage of good-quality land to farm and, especially since the mid-1970s, could be seen as a response to loss of production during the drought years. However, as Lericollais (1989) notes, it also generally reflects a deliberate decision based on a family strategy to cope with environmental risks that are seen as inescapable and largely unchanging. Emigration may also play a cultural function in the transition to manhood, or be economically linked with generating sufficient revenue to buy livestock (USAID, 1990). Factors such as the decline of markets for traditional cash crops (e.g. gum arabic and cotton), and the development of the groundnut basin of Senegal and subsequently of mechanised agriculture in the delta, provide additional motivations to move (see Adams, 1977).

Similar complexity surrounds the 'natural' environment of the Senegal River valley, which has been influenced in recent years both by 'natural' cycles of climatic and vegetational change and by human impact unrelated to the refugees' presence. Thus successive droughts which affected the whole of the Sahel region, particularly in the 1970s and early 1980s, were also characteristic of the middle valley, and are widely interpreted as being part of a process of 'desertification' of an already fragile ecological zone. Deforestation in particular is a major concern, exacerbated by the impact of regulation of the flow of the Senegal River. According to Betlem (1988), between 1954 and 1986, the area of forest reserves in the département of Podor, which has the largest area of such reserves in the country, decreased by 70 per cent, with degradation proceeding at the same speed both inside and outside the reserves (Betlem, 1988). For example, the species *Acacia nilotica*, which existed in low-lying areas that were once seasonally flooded (known as the *walo*), declined after the control of floods removed the conditions it requires to survive (Toussaint et al., 1994). Excessive exploitation of forests has also been stimulated by the increase in demand for wood and charcoal in the urban centres, particularly the cities of Saint-Louis, Thiès and Dakar. In 1980, the département of Podor furnished 20 per cent of the nationally controlled production of firewood and more than 25 per cent of charcoal (Daffe et al., 1991). In addition, overgrazing and the shortage of pastures due to drought has also in recent years led to a shift into the floodplain of the river itself, where pasture and forage are more abundant, especially during the dry season (van Lavieren and van Wetten, 1990).

In such a context of long-term environmental change and population movement – not to mention changes in economic conditions and policies which impact on natural resource management – it is clearly difficult to isolate the impact of refugees on their host environment. This is borne out by a review of changes in vegetative cover, based on analysis of air photographs and satellite imagery from before and after the refugee influx, which provides evidence of contradictory trends in different vegetation types, and a far from clear reduction in overall biomass (Black and Sessay, 1997a). A comparison was made between the broad land cover types existing in 1980 and 1991 within a 10 km radius around each of three villages studied (Figure 6.1). Results of this comparison are shown in Tables 6.1–6.3, separated where appropriate for the *walo* – the floodplain

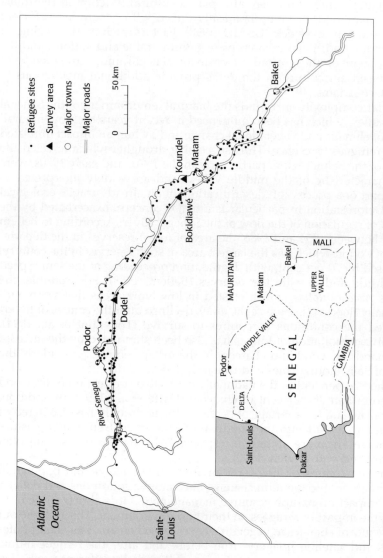

Figure 6.1 Location of study sites in middle Senegal River valley

Table 6.1 Changes in vegetation and land cover around Koundel, Senegal, 1980–91

	1980		1991		Change 1980–91 (%)
	Area (km²)	%	Area (km²)	%	
Walo (floodplain)					
Herbaceous	6.7	7.2	8.5	9.1	+ 1.9
Shrubs	30.4	32.7	24.0	25.8	− 6.9
Shrubs and trees	20.7	22.2	29.0	31.3	+ 9.1
Trees (woodland)	25.8	27.7	8.7	9.4	− 18.3
Forest	2.2	2.4	0.3	0.3	− 2.1
Rain-fed crops	5.7	6.1	4.4	4.7	− 1.4
Irrigated crops	0.8	0.9	8.6	9.2	+ 8.3
Bare ground	0.7	0.8	9.5	10.2	+ 9.4
Total	93.0	100.0	93.0	100.0	

Source: Black and Sessay (1997a).

of the river itself – and the *diéri*, an area of slightly higher land which is not flooded or irrigated, and on which only rain-fed cultivation is possible during the short wet season.[1]

In the village of Koundel (Table 6.1), situated on the banks of the river and close to the Diamel forest reserve, the pattern of change in vegetation and land cover is particularly interesting. The most striking change was in the area under woodland, which decreased from 28 per cent to just 9 per cent of the area studied over the 11-year period, whilst the area classified as 'forest' virtually disappeared, although the former at least may reflect drier conditions during January, when the 1991 image was taken. More certain, however, was a large increase in irrigated cultivation, and in the proportion of bare ground. The process of loss of woodland appeared to have continued up to 1995, although there was some abandonment of irrigated perimeters in the *walo*, as subsidies on fertilisers and breakdowns of water pumps discouraged farmers from continuing irrigated cultivation.

Around the village of Bokidiawé (Table 6.2), there was also a decrease in the small area of woodland remaining, and in the area covered by shrubs, although the category 'shrubs and trees' increased slightly in area, again possibly reflecting a dying away of the woody vegetation during the dry season. However, once again, there was an appreciable increase in the area under cultivation (especially irrigated crops), as well as in the area devoid of vegetation and therefore basically degraded. A closer examination of the results reveals that increases in bare ground occurred almost exclusively in the *walo*, a surprising finding, given the higher agronomic potential of *walo* land. In contrast, on the *diéri*, the area covered by woodland, shrubs and trees increased slightly. As in Koundel,

Table 6.2 Changes in vegetation and land cover around Bokidiawé, Senegal, 1980–91

	1980		1991		Change 1980–91 (%)
	Area (km²)	%	Area (km²)	%	
Diéri (upland)					
Shrubs	60.6	37.4	49.4	30.4	− 7.0
Shrubs and trees	12.9	8.0	17.1	10.5	+ 2.5
Trees (woodland)	0.3	0.2	1.6	1.0	+ 0.8
Rain-fed crops	55.9	34.5	61.0	37.6	+ 3.1
Bare ground	32.3	19.9	33.4	20.5	+ 0.6
Total	162.0	100.0	162.5	100.0	
Walo (floodplain)					
Herbaceous	28.6	20.0	22.5	15.8	− 4.2
Shrubs	64.2	44.9	54.9	38.4	− 6.5
Shrubs and trees	31.8	22.2	29.7	20.8	− 1.4
Trees (woodland)	2.7	1.9	−	−	− 1.9
Rain-fed crops	10.5	7.4	14.8	10.3	+ 2.9
Irrigated crops	1.3	0.9	7.5	5.3	+ 4.4
Bare ground	3.8	2.7	13.5	9.4	+ 6.7
Total	142.9	100.0	142.9	100.0	
Diéri and *Walo*					
Herbaceous	28.6	9.3	22.5	7.4	− 1.9
Shrubs	124.8	40.9	104.3	34.2	− 6.7
Shrubs and trees	44.7	14.6	46.7	15.3	+ 0.7
Trees (woodland)	3.0	1.0	1.6	0.4	− 0.6
Rain-fed crops	66.4	21.8	75.8	24.8	+ 3.0
Irrigated crops	1.3	0.5	7.5	2.5	+ 2.0
Bare ground	36.1	11.9	46.9	15.4	+ 3.5
Total	304.9	100.0	305.4	100.0	

Source: Black and Sessay (1997a).

there was again a retreat from irrigated cultivation in the *walo* during the 1980s.

The pattern of change in the area surrounding the village of Dodel (Table 6.3) shows a similar pattern, with losses in the small area of forest and woodland, and again, increases in the area under cultivation. However, in this case, although the area of bare ground on the *diéri* increased considerably, it reduced by more than half in the *walo*, with increases instead in the area of shrubs and trees. By 1995, relatively few areas of bare ground were visible around the village of Dodel and the adjacent

Table 6.3 Changes in vegetation and land cover around Dodel, Senegal, 1980–91

	1980		1991		Change 1980–91 (%)
	Area (km²)	%	Area (km²)	%	
Diéri (upland)					
Shrubs	9.9	6.5	26.9	17.7	+ 11.2
Shrubs and trees	95.9	62.8	72.0	47.3	− 15.5
Trees (woodland)	2.5	1.6	–	–	− 1.6
Rain-fed crops	44.6	29.1	45.1	29.7	+ 0.6
Bare ground	–	–	8.1	5.3	+ 5.3
Total	152.9	100.0	152.1	100.0	
Walo (floodplain)					
Herbaceous	50.8	32.1	37.8	23.8	− 8.3
Shrubs	40.4	25.5	59.1	37.2	+ 11.7
Shrubs and trees	34.6	21.8	37.3	23.5	+ 1.7
Trees (woodland)	12.5	7.9	4.2	2.6	− 5.3
Forest	1.3	0.8	–	–	− 0.8
Rain-fed crops	4.8	3.0	5.9	3.7	+ 0.7
Irrigated crops	5.3	3.4	10.7	6.7	+ 3.3
Bare ground	8.6	5.5	3.9	2.5	− 3.0
Total	158.3	100.0	158.9	100.0	
Diéri and *Walo*					
Herbaceous	50.8	16.3	37.8	12.2	− 4.1
Shrubs	50.3	16.2	86.0	27.7	+ 11.5
Shrubs and trees	130.5	41.9	109.3	35.1	− 6.8
Trees (woodland)	15.0	4.8	4.2	1.4	− 3.4
Forest	1.3	0.4	–	–	− 0.4
Rain-fed crops	49.4	15.9	51.0	16.4	+ 0.5
Irrigated crops	5.3	1.7	10.7	3.4	+ 1.7
Bare ground	8.6	2.8	12.0	3.8	+ 1.0
Total	311.2	100.0	311.0	100.0	

Source: Black and Sessay (1997a).

refugee site, with relatively dense cultivation in the *walo* in particular. Part of the bare ground recorded on the *diéri* in 1991 was certainly occupied by the refugee site itself, which was one of the largest in the middle valley.[2]

Such an analysis has many limitations – not least the availability of remotely sensed imagery that captures land use change from the pre- to post-refugee period. However, what it shows is hardly a progressive process of deforestation and desertification; a view supported by interviews with local people and refugees themselves. These interviews established that a secular decline in vegetation was not identified by respondents since the refugees' arrival (although the decimation of crops and herds during the earlier drought was seen as an issue). It also showed no evidence at all of refugees using natural resources – whether woodfuel or agricultural land – in a different way to that of local populations, with the exception of a slightly lower level of access to irrigated land amongst the refugee population. Thus cultivation practices were similar; energy consumption figures showed no significant difference; and attitudes towards the use of different species for woodfuel were largely the same (Black and Sessay, 1997b).

The Forest Region of Guinea

The refugee influx into the Forest Region of Guinea from 1989 onwards was both larger and longer-lasting than in the Senegalese example cited above. Thus after an initial influx from Liberia, subsequent arrivals included significant numbers in 1990, 1993 and again in 1995 as the war intensified first in Liberia and then in neighbouring Sierra Leone. Unlike the Senegal River valley, at the time of the first exodus of refugees, few international agencies were operating in the Forest Region of Guinea. In the absence of any significant international programme of assistance, refugees settled in local villages by arrangement with local communities (van Damme, 1997). By 1995, it was estimated by the local field office of the UNHCR that the 81 000 refugees in Yomou préfecture were settled in 87 different villages, 27 of which received more than 1000 refugees. All of these sites, except one, were located on the fringe of existing population settlements; the exception, Nonah, was a new camp set up in 1994 in a relatively underpopulated area on the edge of the Diécké Forest Reserve (Figure 6.2).

However, as in Senegal, food assistance was eventually provided by the international community; whilst similarly, by 1995, the UNHCR had begun a policy of phasing out food rations for at least some refugees.[3] In addition to the dispersed nature of settlement, a second key element of the refugee presence in Guinea has been the form in which international assistance has been provided, which again has some similarities with Senegal. In particular, although food aid has been delivered only to refugees, other elements of emergency assistance, notably health care and water, have been provided through government ministries, and were targeted at both refugees and local communities. Thus in providing health care, rather than constructing separate clinics and hospitals for refugees, the UNHCR and other international agencies, notably MSF-Belgium, sought to support existing health services so that they could cope with an increased case load (van Damme, 1997). Similarly, education was provided by doubling-up classes in existing school buildings (although a separate education system for refugees in English has operated), whilst wells were dug for clean water in both refugee sites and local villages.[4]

Although less 'open' to the outside world than the Senegal River valley, particularly since independence from France in 1958, analysis of the

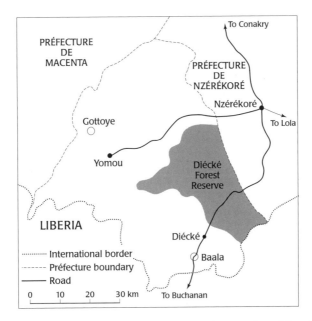

Figure 6.2   Location of study sites in Yomou Préfecture, Republic of Guinea

recent history of the Forest Region of Guinea also provides an important context of environmental, demographic and especially socio-political change, which is essential to an understanding of the impacts 'caused' by the presence of refugees. Thus in the past, there has been a dynamic regional process of population movements, beginning (in known history at least) with the migration of the Mande-speaking peoples from the West African savanna to the forest in the period from AD 1350 to 1500. Germain (1984) suggests that significant movement of the Kpelle into the region from the north occurred from the 16th and 17th centuries onwards, whilst Nelson et al. (1975: 70) describe the Kpelle as settling in the region at that time to avoid being troubled by the Mandinka. The Konianké are more recent arrivals, with significant movements even in the post-independence period, mainly of traders. At the time of French colonisation (around 1905–1907), the border between the Kpelle and Mano areas was almost certainly in a state of flux as the two groups fought for control. Conflicts were probably exacerbated by the presence of competing European interests in Guinea and Liberia (Germain, 1984).

The Forest Region was colonised late by the French and the period of colonial rule was relatively short, ending abruptly in 1958 with the coming to power of President Ahmed Sekou Touré. During the colonial period, the population of the Forest Region appears to have remained relatively stable, although there was probably some out-migration to major cities such as the capital, Conakry. After independence, however, there was significant movement out of the Forest Region, both to Conakry and, especially, to Liberia and Côte d'Ivoire. This latter movement represented a response to declining economic conditions and strict political controls in Guinea, as the country became a socialist state completely isolated from the rest of Francophone West Africa. It also reflected the opportunities presented by the rapidly developing economies of Guinea's neighbours.

Whilst evidence on recent environmental change in the particular case of Yomou Préfecture is scarce, some evidence is provided by recent work by Fairhead and Leach (1996) in the préfecture of Kissidougou in the forest–savanna transition zone to the north. Thus the standard 'narrative' of environmental change, and the most serious environmental concern in Yomou, is that of deforestation. Grainger (1993) estimates that West Africa as a whole has already lost 70 per cent of its 'climax' forest, and accounts for over half of total African deforestation, although the period over which this has occurred is not given. Rates of deforestation are not provided by Grainger for Guinea, but concern is reflected in country-level reports. For example, Bourque and Wilson (1990: 121) argue that *'Guinée Forestière* was originally, as its name implies, wholly forested', and they go on to state that 'deforestation is extensive, and going on at a rapid pace. The general consensus is that all natural forest outside the forest reserves, excepting fragments protected by terrain or tradition, is liable to be lost in the foreseeable future' (Bourque and Wilson, 1990: 122).

However, Fairhead and Leach (1994: 483) contest this narrative, arguing from evidence in Kissidougou that:

> Although intensive human use has often been blamed for the savanniza-tion of forest, it can have the opposite effect; past or present land-use practices can enable forest to develop in savanna where otherwise this would be unlikely for want of the necessary conjuncture of particular soil moisture, fire limitation and seed conditions.

Indeed, Fairhead and Leach contest what they see as the dominant narrat-ive across Africa as a whole, in which 'original' forests that represented a 'climax' vegetation have been degraded after a breakdown of the tradi-tional 'functional social organisation', as a result of either population in-creases or immigration, or both (Fairhead and Leach, 1995: 1032). Citing the specific examples not only from Kissidougou Préfecture but also from the Ziama Forest Reserve, around which a number of refugees were settled, they challenge both the 'fact' of deforestation in Guinea and the narrative that surrounds it, in which the environment is seen as unchanging except by virtue of (negative) human influence. Powerful evidence for their argu-ment that the area of forest has actually increased rather than decreased in historical times, is provided by documents, oral histories and satellite imagery – although this evidence currently does not extend to Yomou Préfecture itself.

Evidence for environmental change in the areas directly affected by refugees in Guinea suffers from the same difficulties of distinguishing it from longer-term and sometimes contradictory trends and contexts cited for Senegal above, and in this case is hampered by a lack of available air photograph and satellite images for the region. Certainly there have been

some impacts; for example, Sawyer (1990) cites serious pressure of encroachment by Liberian refugees on the Mont Nimba Reserve in the far
east of the Forest Region, in an area designated as a World Heritage Site
and Biosphere Reserve (IUCN, 1991). Similar concerns are expressed by
Bourque and Wilson (1990) about the Ziama Reserve, also a Biosphere
Reserve, in the préfecture of Macenta to the west. The real extent of this
pressure remains unclear though, with Mont Nimba at least currently
suffering in any case from a lack of protection even from local populations
as government plans to open up the zone to iron ore mining after the end
of the Liberian war have brought a halt to most conservation efforts.[5]

A review of impacts in Yomou Préfecture suggests that three different
types of potential impact of refugees on environmental change need to
be distinguished: first, the complete removal of forest cover leaving bare
ground; second, a conversion of forest to agricultural land; and third,
pressure on particular resources and species within the forest itself (Black
and Sessay, 1997c). Concerning complete removal of forest cover, there
is little evidence that this has happened on any great scale in Guinea,
although within some refugee sites the clearance of trees had led to some
gullying and soil erosion on a localised scale. More significant is pressure
on particular forest flora and fauna, with refugees seeking a variety of
products including wood for fuel and shelter, raffia (for thatching), palm
branches (for thatching), nuts (for making 'wine' and soap), and 'forest
foods' such as wild yams, mushrooms and snails. It was also clear from
field surveys that refugees had generally been forced to use 'lower quality'
alternative products in situations where resources were not abundant –
such as palm rather than raffia for thatching – whilst they have not had
access to some resources (e.g. timber) at all, except through purchase
from locals.

However, evidence is presented in Table 6.4, which shows that the
income-generating activities conducted by refugees were much less reliant on natural resources than those conducted by locals. For refugees, the
main activity other than farming was contract work, either for local villagers or on newly established rubber plantations in one of the study
villages. Significant numbers were also involved in commerce and 'professional' activities (mainly teachers or camp leaders, although remuneration
was often low), where environmental 'impacts' were either minimal or
non-existent. In contrast, for local households, all of the major activities
identified relied to a greater or lesser extent on the use of natural resources.
Thus fishing, palm nut collection and the production of raffia wine (by
men) and the processing of palm oil and other forms of craft production
(by women) were amongst the most important activities. In each case,
the access of refugees to these natural resources was much more limited.
The production of palm kernel oil by refugee women forms an interesting
exception to this pattern, since this is seen by locals as a resource to be
used only in times of emergency cash need (*cf.* Falconer, 1990). At other
times, the cost of production (especially in women's labour) is seen as
outweighing the value of the final product.

Perhaps the most significant impact of the refugees in the Forest Region
of Guinea was their cultivation of agricultural land, which has certainly

Table 6.4 Income-generating activities for refugees and locals in three study villages in the Forest Region of Guinea (with percentages in parentheses)

Activity	No. of refugees	No. of Guineans
Contract work	48 (40)	1 (1)
Commerce	25 (21)	11 (12)
Fishing[1]	20 (17)	44 (48)
Work for Soguipah[2]	20 (17)	4 (4)
Palm kernel oil	19 (16)	8 (9)
Professional[3]	17 (14)	3 (3)
Palm nut collection	14 (12)	29 (32)
Sale of crops	10 (8)	13 (14)
Construction	6 (5)	8 (9)
Craft production	6 (5)	11 (12)
Sale of wood	4 (3)	1 (1)
Sale of rice	3 (3)	4 (4)
Blacksmith	2 (2)	3 (3)
Charcoal production	2 (2)	1 (1)
Raffia wine	– (–)	23 (25)
Total households	120 (100)	92 (100)

Notes:
[1] Does not necessarily involve sale of fish.
[2] Soguipah is an industrial oil palm and rubber plantation company.
[3] Includes teachers, nurses, and those with remunerated administrative responsibilities in refugee camps or local villages.
Source: Black et al. (1996).

placed pressure on remaining forest resources. However, even here, there seems to be less a process of conversion of forest to agricultural land than an intensification of cultivation on land that has already been cultivated. The main process by which this has taken place has been a shortening of the fallow cycle on 'uplands'[6] within what is broadly a 'bush fallow' (or 'slash and burn') system. Parallel to processes noted in Chapter 2, refugees often cultivate land that has been fallow for only three (or fewer) years rather than the seven or more years that was common in many areas before their arrival. Such land is not only less fertile, but its cultivation can lead to further reduction in fertility and especially to weed encroachment, reducing future yields. None the less, a number of processes are operating here. First, there appears to have been a reluctance either of refugees to cultivate 'stronger bush' (land which has been left for a longer fallow) or of locals to allocate such land to refugees. In addition, a common occurrence was found to be the granting of land to refugees for one season only,[7] after which the 'owner' would plant coffee or cocoa on the land that had been cleared. This represents a cheap method for local farmers to open up new coffee plantations, which in 1995 was a major goal for many local people in the wake of increases in coffee prices, and a relaxation of strict government controls on coffee marketing that had been introduced during the Sekou Touré era (and at the time had led to many in Yomou ripping out their coffee plantations).

Agency strategies: the search for a sustainable system

The picture painted above is one of dynamic change in vegetative cover, and particularly an expansion of agricultural crops, but the extent to which this can be described as 'environmental degradation', or even permanent environmental change, can be called into question. In both cases, by 1995, donors were moving towards withdrawal of food rations as implementing agencies promoted what appeared to be successful cases of refugee 'self-sufficiency'. In this context, it is interesting to ask why, despite the apparent fragility of the environment in both Senegal and Guinea, the effect of large-scale influxes of refugees was not to massively increase degradation of resources, and why in some cases, the changes occurring might even be seen as 'beneficial' in terms of increasing agricultural production.

There are several potential answers to these questions, relating to both the activities of external assistance agencies (whether international organisations or the national government of each country) and refugees and local people themselves. In terms of the actions of external agencies, policies at two levels were especially influential in affecting the patterns of natural resource use and vegetation change observed in the two refugee-affected areas. Thus in both cases, the pattern of settlement of the refugees was highly dispersed (although this was less the result of policy and more a *fait accompli* in Guinea), and it could be argued that this led to a dispersal of the pressure on natural resources – unlike the situation in the Rwanda emergency nearly five years later. The role of assistance policies, which focused on supporting the refugee-affected area rather than the refugees themselves, could also be seen as influential in many respects, by creating (or supporting) conditions in which refugees and local populations both had a stake in working together and co-operating over management of natural resources in order to benefit jointly from any assistance provided from outside.

In addition to this broader policy level, there was also specific policy in the two countries towards management of land resources, although only in Guinea did this involve overtly 'environmental' projects on the part of international agencies assisting the refugees. Thus in Senegal, one major intervention of the UNHCR was to support refugee agriculture on irrigable land in the valley bottoms, through assistance in the creation of *'perimètres irriguées villagoises'* (PIVs). Such schemes had initially been developed by the state in the late 1970s and early 1980s, through the Société d'Aménagement et d'Exploitation des Terres du Delta et du Fleuve Sénégal (SAED), in order to establish a basis for the development and management of irrigated land by groups of 30 to 50 farmers (Figure 6.3). Each farmer in the group would cultivate a small plot and share a small motor pump, on land prepared by SAED;[8] the necessary equipment (motor pumps and accessories), and inputs (seeds, fertilisers, pesticides)

Figure 6.3 Irrigation scheme for refugees and local people, Matam, Senegal
Source: (photo: Richard Black)

were provided by SAED at highly subsidised prices, so that often 50–70 per cent of production costs were subsidised. Easy access to credit was also provided to farmers within the scheme, with recovery in principle over a short term of two to three seasons and often in the form of crop (seed grain) after harvest.

Prior to the arrival of the refugees, subsidies for such schemes had been phased out as part of a structural adjustment package which saw a disengagement of the state from the agricultural sector (Woodhouse and Ndiaye, 1990). Much irrigable land in the valley remained outside these schemes, whilst some PIVs required redevelopment as they had fallen into disuse. The UNHCR scheme provided for renewed subsidy of irrigation development, and provided for mixed groups of refugees and locals to benefit from assistance (Mbodj et al., 1995). Meanwhile, a separate initiative in Senegal involved the establishment of village forestry committees, and broader village natural resource committees, through the action of the state Forestry Service. Although not overtly aimed at refugees, by 1995 it was common for refugees to be represented on these committees, which were nominally responsible for managing some land in both forest reserves and village forests. Interviews conducted at village level in Senegal suggest that these committees were viewed by refugees and locals alike as still largely in the control of the state Forestry Service. None the less, in some cases they were able to enforce certain rules on tree cutting and use of woodland, and at the very least, they represented a symbolic incorporation of refugees into a formal institution concerned with land management.

In Guinea, agricultural projects involving the development of irrigation schemes had a more overtly 'environmental' purpose for the UNHCR

– perhaps surprisingly given the much greater resilience and regenerative capacity of ecosystems in the Forest Region – whilst forestry schemes were also funded in order to address environmental concerns. Thus in addition to seeking to promote 'self-sufficiency', the UNHCR's support to agricultural projects also sought to address the question of deforestation by providing an opportunity to intensify production on seasonally flooded swamps. With good management, improved varieties and some use of chemical fertilisers, it was argued that such swamp lands, which were generally left uncultivated in the traditional system, could produce yields three to four times higher than the surrounding uplands. In most cases, it was also theoretically possible (as in Senegal) to obtain two or even three rice harvests each year, against only one on the uplands. Clearly, the UNHCR and the government argued, if refugees (and local populations) could be given an incentive to cultivate in the swamps rather than on the uplands, agricultural production and productivity would rise considerably, whilst pressure on remaining upland forests would be released.

From small beginnings in 1991, schemes for the drainage and management of swamps funded by the UNHCR grew over the 1990s, such that by 1996, a total of 507 ha of swamp land had been 'developed' across the Forest Region (SNSA, 1996). This 'development' took two forms: first, the simple drainage, levelling and division of the swamps into small plots (typically 0.5 ha per family); and second, the above measures plus the construction of sluice gates in the water course to control the water level in the swamp. As in Senegal, these schemes were inspired by existing government initiatives to develop swamp land, although unlike Senegal, these existing initiatives, funded by the World Bank, IFAD and the French government amongst others, were still ongoing as the UNHCR projects were implemented (Figure 6.4). In addition, the UNHCR projects again involved mixed groups of refugees and locals in the development of swamps, with short-term guarantees of access to land for refugees being 'traded' for long-term investment in swamp land owned by local families and communities.

In both Senegal and Guinea, relatively small-scale irrigation schemes promoted by the UNHCR played a role in encouraging intensification of agriculture, and demonstrated that it was possible for refugees and local people to work together on the basis of shared (though distinctive) interests in the development of agricultural projects. In Guinea in particular, it was possible for the UNHCR to 'integrate' its interventions into the broader agricultural development strategy of the government, working through the same operational partners (principally ministries of the Guinean government) to address both agricultural productivity and environmental concerns. Despite these positive features, the schemes cannot however be regarded as purely 'beneficial' from an environmental point of view. First, reticence on the part of local populations in Guinea to develop swamps for intensive rice cultivation reflected the high value of products already derived from the swamps, notably raffia products. In this case, the development of the swamps involved a trade-off between these products and rice production. In addition, the levelling of the swamps could be seen as contributing to the destruction of the 'biodiversity' of the area, in the sense that they

Figure 6.4 Swamp development scheme in Baala, Republic of Guinea
Source: (photo: Richard Black)

represent important and diverse wetland areas. The reliance of swamp development schemes on chemical fertilisers to increase productivity might also be questioned from an environmental point of view, although one local (Liberian) NGO at least was experimenting with intensive swamp cultivation using 'organic' techniques and green manuring.

In addition, both the UNHCR swamps, and swampland developed as a result of intervention by other donors, have generally failed either to produce the yields initially expected or to be cultivated for more than one cropping season per year. A survey by the Guinean Service National de Statistiques Agricoles (SNSA, 1996) showed average production figures on UNHCR-developed swamps at only 1.6 t ha^{-1}, the lowest of seven projects analysed nation-wide, and well below the level of 3.5 t ha^{-1} initially expected in project documents. Many refugees and locals who farmed in the newly developed swamps continued to cultivate an upland farm, since they were able to grow a range of associated crops (cassava, beans, maize, etc.) and obtain an earlier rice harvest than on swamps solely devoted to paddy rice. In addition, there appears to have been a decline in productivity on swamps of all kinds after two to three years of cultivation as a result of varied problems, including poor management and extension advice, sand deposition, iron toxicity and the flushing of nutrients away from the plots in irrigation channels. It is interesting to note that similar problems with intensive swamp rice development had been observed in the 1950s and 1960s in neighbouring Liberia and Sierra Leone (Binns, 1982).

Indigenous strategies and initiatives: informal regulation

The interventions of the UNHCR and other international agencies in Senegal and Guinea provide a partial explanation for both good relations between refugees and their local hosts, and the creation of conditions in which natural resource management goals could be promoted and accepted by the two populations. None the less, these interventions arguably played only a minor role both in the pattern of natural resource management and in influencing the day-to-day activities of the refugees and local land managers. Rather more important, especially in Guinea, was the fact that procedures for access to land and natural resources had been negotiated by refugees quite independently of these external agencies. As a result, the status of local leaders, and of a range of informal institutions of relevance to natural resource management, had in many cases been strengthened, rather than weakened, by the presence of the refugees.

An important example of this is provided by forest protection, promoted both formally by the state and, in the case of Guinea, informally by local institutions. One project, funded by the UNHCR, and carried out in collaboration with the state Forestry Service, involved reforestation on a total of 240 ha during 1992–95, spread over 20 sites in the Forest Region. Its success though has been limited, not only in terms of area planted but also in terms of its acceptance, and the respect shown for plantations, by refugees and locals. Interviews with local officials of the Forestry Service in 1995 suggested that the majority of the sites involved commercial plantations of pure stands of species such as *Acacia mangium* or *Terminalia superba*, which although partly 'indigenous', hardly addressed concerns of biodiversity in the area. More critically, it was accepted that a number of the sites chosen were, by necessity, previously abandoned forest plantations, controlled by the Forestry Service itself rather than local populations. Limitations imposed by Guinea's forest laws prevented the Forestry Service from encouraging more local participation, for example through conferring ownership of the trees on local communities.

Yet the plantations which were 'renovated' had originally fallen into disuse precisely because of hostility from local populations, who saw them as usurping traditional institutions of land management and ownership. In the 1995 campaign, some local authorities (*communautés rurales*) were persuaded to cede land for new plantations, and a larger range of tree seedlings were planted. However, the laws on ownership remained unchanged, whilst the trees planted still did not 'replace' natural forest in the strictest sense. As property of the state, nor on the whole did the trees provide woodfuel or timber for local purposes. Rather, they can be seen – and were seen locally – as a 'cash crop' to be disposed of by the state to raise state revenues. Whilst the Guinean state undoubtedly incurred expenses in dealing with such a large refugee influx, such plantations were little different conceptually to commercial plantations of rubber, which were being promoted elsewhere in the region, and which were also seen by some as contributing to the loss of tropical rain forest.

Figure 6.5 Refugee site adjacent to forest in Yomou Préfecture, Republic of Guinea *Source*: (photo: Richard Black)

In contrast, many 'sacred forests' which exist in local villages appear to have been well-protected, with local institutions maintaining their authority through a mixture of social pressure and fines for infringement of rules. Prohibitions on resource use in sacred areas in general (e.g. sacred forests, rivers where fishing is prohibited) have been maintained by fines and beatings, whilst the refugees have largely respected community ownership of specific forest resources such as large trees suitable for timber, raffia, and oil palm. The result is that tree cover adjacent to most refugee sites has been maintained (Figure 6.5). Although some petty theft certainly exists, and was vigorously complained about by local leaders during interviews, there also appeared in 1995 to be an acceptance that a low level of theft was perhaps inevitable, and that this had been maintained within reasonable limits. In one sense, the existence of a potential external 'threat' to sacred forests (many of the Liberian refugees are from evangelical Christian sects who reject the animist belief in sacred forests) has provided a basis for those who control them to mobilise local populations in their defence.[9]

What has been crucial to maintaining acceptance of local 'rules' concerning forest resources in both Guinea and Senegal has been the fact that there has been an acceptable level of access for refugees to the natural resources they need to survive, although this access has been by no means complete. Surveys conducted in the two case study areas show that refugees have gained access to land resources (Table 6.5), in the case of Senegal either through family or lineages (or through rental); and in the case of Guinea through working for local people, finding a 'stranger-father', and offering token gifts (a chicken, kola nuts, and usually a small percentage of the rice crop). Of these, the former is more secure, and reflects the close kinship

Table 6.5 Types of cultivation by surveyed refugee and local households in Senegal and Guinea

| | Number of households (%) | | | | | |
| | Refugees | | Local residents | | Total | |
	n	(%)	*n*	(%)	*n*	(%)
Senegal						
Flood recession crops (*walo*)	47	(78)	54	(90)	101	(84)
Rain-fed crops (*diéri*)	57	(95)	48	(80)	105	(88)
Irrigated crops	0	(0)	24	(40)	24	(20)
No farm	2	(3)	4	(7)	6	(5)
Total households	60	(100)	60	(100)	120	(100)
Guinea						
Swamp rice	30	(25)	34	(37)	64	(30)
Upland rice	69	(57)	81	(88)	150	(71)
Cassava on second-year plots	30	(25)	35	(38)	65	(31)
No farm	30	(25)	3	(3)	33	(16)
Total households	120	(100)	92	(100)	212	(100)

Source: Black and Sessay (1998).

ties that have always existed between populations on either side of the international border between Senegal and Mauritania. However, even though the latter is precarious, and has led to a smaller percentage of Liberian refugees in Guinea gaining access to land, it does reflect the fact that local institutional practices already allowed a route through which access to land and other resources can be obtained by strangers. Indeed, such access had been provided in the past to other stranger groups, most notably the Malinké group, which continues to move into the area from the north. Local structures have provided access not only to land but also to other crucial resources such as woodfuel, construction materials and water.

This is not to say that refugees have been fully 'integrated' into local informal natural resource management institutions; indeed, in many cases, they have not been 'integrated' at all, but rather 'managed' by them. In many respects, the state-created village forest committees in Senegal, where Mauritanian refugees achieved some representation, have been the exception rather than the rule. Elsewhere, a number of informal and locally run institutions in both countries have not been opened up to refugees. In Guinea, for example, the '*Association des Sages*', or committee of village elders (usually men), represents the founding lineage or lineages of the village, and oversees land disputes. This institution, whose membership also overlaps with the control of secret societies (who are responsible for sacred forests), has not involved refugees in decision making, despite being a powerful force in making decisions on issues of land and natural resources. Similarly, in Senegal, the *'jom leydi'* or 'guardian of the land', who in principle arbitrates the competing demands of farmers, pastoralists and fishermen (Salem-Murdock et al., 1989), has remained in local hands.

More importantly, such local informal institutions do not necessarily operate in a way that is equitable to all concerned, nor are they divorced from questions of economic and political interests of individuals and groups who control local (and sometimes national) institutions. In both Senegal and Guinea, there is clear evidence that land made available to refugees has been, on balance, of poorer quality, reflected mainly in the higher use by refugees of land outside the floodplain in Senegal, and the allocation of land that has experienced a much shorter fallow cycle to refugees in Guinea. Local institutions have allowed refugees access to land in the short term, but face a much more severe test if this is to be extended into the future. This situation is hardly surprising: indeed, it is quite clear that such local institutions are also not inclusive of all of the members of host societies either, with questions over both social and gender equity in the control and distribution of resources.

It is also worth noting that one very effective form of 'institutional' regulation of natural resource use in Senegal has been the market – in particular in the case of woodfuel use. The very low rates of woodfuel use found in the Senegal River valley amongst both refugee and local populations (at less than 1 kg per person per day) can be partly attributed to the fact that wood has to be purchased by between a third and one-half of all refugee and local women (Black and Sessay, 1997b). Given the scarcity of woodfuel, its purchase price is relatively high, and a system has developed to supply it from neighbouring areas (Figure 6.6). As a result, there is a powerful incentive for both communities to economise in their use of wood, although clearly there is again an equity question concerning the households which can and cannot afford to purchase wood for fuel.

Local institutions and environmental management

The cases of Guinea and Senegal provide two examples of the role that can be played by local informal institutions in natural resource management, even in situations of mass forced migration. Far from leading to the collapse of local institutions, the case studies provided in this chapter suggest that large-scale in-migration of refugees can act to stimulate such institutions as local leaders find it in their economic and political interest to provide certain rights to refugees, effectively taking them on as political 'clients'. Of course, such a situation is perhaps easier to envisage in a context where refugees become relatively dispersed amongst the local population, so that numbers in any one village are not so large that they overwhelm existing management institutions. In addition, the existence of ties of 'kinship' between refugees and local people might be cited in both cases as contributing to relatively harmonious relations between the two populations, and collaboration in the field of natural resource management. However, the importance of kinship can be overstated for Guinea,

Figure 6.6 Woodfuel supply, Matam, Senegal
Source: (photo: Richard Black)

where a significant minority of the refugees in each area were not from the 'dominant' ethnic group, whilst the very different histories of the two countries had led to important differentiation between members of the same ethnic group on either side of an artificial colonial border. Perception of the potential advantage to be gained through political patronage may be also a force that encourages the dispersal of refugees in host societies, as appears to have been the case in Bukavu in eastern Zaire in late 1994. Moreover, even in Guinea and Senegal, the number of refugees relative to the local population in any one village was often very high, with many examples of villages in which there were two or three times as many refugees as existing local inhabitants.

Meanwhile, despite a number of distinctive features, the cases of Senegal and Guinea are by no means unique in refugee situations in terms of the potential role that can be played by local institutions. For example, the World Bank's (1996a) evaluation of the 'IGPRA' project for areas of Pakistan affected by Afghan refugees notes how a local institution – the *jirga* – could have fulfilled an important function in ensuring the sustainability of reforestation projects in the region, but was ignored in favour of the creation of a 'modern' village development committee. The *jirga* is not a permanent entity, and can take various forms; what is important is that it provides for local decision making on the basis of consensus, and despite no mechanism for enforcement, it enjoys widespread acceptance. The bank's conclusion was that the creation of village development committees was not only unnecessary but may have undermined the sustainability of forestry projects, by creating greater dependence on

149

the Forestry Department. In addition, whilst by no means an overtly equitable institution, the *jirga* was seen – too late – as having the potential for being more inclusive, particularly of the poor, than externally created committees could ever have been.

Examples of successful, or potentially successful, management of natural resources by local institutions are important, but it is also clearly important not to 'romanticise' the role that they can play in refugee-affected areas, or indeed in any other part of the rural 'south'. As is examined in more detail in the next chapter on post-conflict situations where refugees return to their 'home' societies, there are a number of important limitations on such community-based initiatives, in particular concerning their level of inclusiveness in defining who is considered a part of the 'community'. For refugees, the lack of representation in local informal institutions is perhaps not surprising, given their status as outsiders in the zones in which they are living. In such situations, what appears to be important is the extent to which local institutions are capable of adapting to respond to rapid population increase, and of allowing, albeit to a limited extent, some level of access to natural resources for those who have newly arrived in the area. The evidence from Guinea and Senegal at least is that such local-level institutions may be far from dead, and can play a constructive role in ensuring that the impact of refugees is not a 'disaster' for local environments. The challenge for external agencies wishing to assist refugees is to engage with such structures, to ensure that positive benefits for refugees, and a more 'sustainable' pattern of natural resource use, can be achieved.

1. For further details on methodology and analysis of results, see Black and Sessay (1997a).

2. The Dodel refugee site had a total population of just over 2000 according to local UNHCR statistics – hardly large by global standards!

3. This has been a complicated process, since there were serious interruptions in the supply of food to all refugees during the rainy season in 1995.

4. In Senegal, separate facilities were provided for refugees in some cases, although there has been some effort to provide an integrated programme of assistance to the 'refugee-affected area', albeit on a smaller scale than in Guinea.

5. It is interesting to note that the war in Liberia has held up iron ore extraction on the Guinean side of the Mont Nimba massif for at least seven years, since it is too expensive to ship the ore over 1000 km to Conakry. However, with peace in Liberia, it seems likely that this resource will be exploited through open-cast mining and shipped to the port of Buchanan, as had already happened, with devastating environmental consequences, on the Liberian side of the massif. For a somewhat dated review of threats to the Mont Nimba reserve, see Lamotte (1983).

6. This region of Guinea can be divided into uplands and seasonally flooded swamp areas. Both are suitable for rice cultivation, although using different varieties of rice, and involving radically different production systems. Upland cultivation is more widespread, and plays a greater role in the 'traditional' agricultural system.

7. In some cases, an additional crop of cassava would be allowed, although cassava is seen as a drought crop and is not a major part of the diet except in extreme circumstances.

8. Land preparation included the levelling of plots, the digging of canals, and the intro-duction of a water management system.

9. There has also been a more general strengthening of animist religious practices in the Forest Region in recent years, in part as a backlash after severe repression under the Sekou Touré government.

CHAPTER 7

Post-conflict environmental rehabilitation

In this chapter, the focus of discussion is turned to the repatriation of refugees. On the one hand, repatriation might be seen in part as linked to environmental damage caused by refugees' presence in host areas. Once repatriation has started, the question arises as to how quickly degraded host areas might 'regenerate', and whether any policy interventions are required to assist in 'rehabilitation'. In addition, the repatriation of refugees to their home countries also raises a number of new resource use questions associated with 'population pressure' that are similar to those in countries of asylum. Perhaps the most significant difference is that on return to their 'home' countries, refugees find themselves once again in a country where they are citizens rather than 'strangers'. As a result, their participation in political and economic structures which affect environmental management is potentially part of a process of national reconstruction, rather than something to be achieved in competition with local 'owners' of the resource. Drawing on case studies from a range of countries, the respective roles of international agencies and local institutions is once again called into question in this chapter, and the resilience of local solutions to environmental management problems is explored.

Issues in post-conflict environmental rehabilitation

Early in her term of office, the current United Nations High Commissioner for Refugees, Sadako Ogota, made the well-publicised statement that the 1990s were to be the 'decade of repatriation'. Although there have been some notable failures of some long hoped-for returns to occur, the figures presented in Chapter 1 of declining numbers of refugees over the last two to three years bear some witness to the repatriations that have occurred. For the UNHCR, voluntary repatriation is seen as the most desirable 'durable solution' for refugee crises, implying a homecoming for those forced to flee their countries by persecution, war and armed conflict, and an opportunity to rebuild their livelihoods in their place of origin. As such, of all the large-scale population movements associated with forced population displacement, repatriation represents perhaps the best chance in which such movements could be linked to 'sustainable development'.

But does the repatriation of refugees represent a 'durable solution' for the environment? In former refugee-hosting areas, is repatriation a sufficient condition for environmental recovery and rehabilitation, always assuming that the impact of the refugees' presence was negative in the first place? And is the reintegration of refugees in their home countries and places of origin any less complex than the initial outflow of refugees to places that are not their 'homes'? In these questions lie broader clues as to whether the category of 'refugee' or 'forced migrant' has any significant analytical power in explaining links between people and their environments – or whether it is simply the phenomenon of migration (or even of social and economic change) that is most important, with the legal and social status of individuals playing only a minor role. At the same time, if environmental management issues prove to be similar in former refugee-affected or returnee areas to those in countries of asylum, this could provide some support to the notion that refugee situations are not so 'unusual', and that some inferences at least can be drawn from their experience for more general questions concerning environmental management and sustainable development.

An initial issue that is thrown up by repatriation of refugees is whether the increase in the number of people returning to their countries and places of origin has any connection with environmental damage caused in host areas in the first place. In other words, have the fears of some been justified, that the environmental impact of refugees would be so severe that 'an area's resources . . . will become unable, in the short or medium term, to support the indigenous population, and thereby create an additional class of environmental refugee' (ERM, 1994: iii). At present, there is little evidence to support such a position. Table 7.1 identifies the large-scale repatriations[1] that have occurred during the 1990s. Somewhat reassuringly, large scale repatriations during this period have most commonly been associated with an end to war and conflict in the country of origin, often linked with UN peace-keeping and peace-building operations and supervision of democratic elections such as in Cambodia and Mozambique; or at least with the establishment of new regimes as in Ethiopia, Eritrea and Rwanda. Even if peace and stability have sometimes been temporary (as in Cambodia), or repatriations have sometimes been forced (as from Rwanda to Burundi in 1994, or arguably from Congo and Tanzania to Rwanda in 1996 and initially from Mexico to Guatemala), there are no clear-cut cases in which environmental degradation in the host area appears to have been either a primary or even an important factor in encouraging return.

Lack of clarity about the situation has not, of course, prevented some at least from suggesting that environmental damage is a reason for refugees to be repatriated. For example, as was argued in Chapter 5, the Tanzanian authorities cited environmental damage 'caused' by refugees

Table 7.1 Mass repatriations in the 1990s

To	From	Year	Numbers
Afghanistan	Pakistan	1990–94	2 079 000
Iraq	Iran	1991	1 300 000
Mozambique	Malawi	1992–94	1 290 000
Rwanda	DRC	1996	1 000 000
Afghanistan	Iran	1990–94	905 000
Rwanda	Tanzania	1996	550 000
Ethiopia/Eritrea	Sudan	1991–95	500 000
Ethiopia	Somalia	1991	500 000
Lebanon	Various	1994	450 000
Iraq	Turkey	1991	448 000
Kuwait	Saudi Arabia	1991	400 000
Burundi	Tanzania/Rwanda	1994	400 000
Guatemala	Mexico	1991–94	300 000
Sudan	Ethiopia	1991	270 000
Cambodia	Thailand	1992	249 000
Somalia	Ethiopia	1991–92	200 000
Mozambique	South Africa	1992–94	200 000
Mozambique	Zimbabwe	1992–94	197 000
Liberia	Côte d'Ivoire/Guinea	1991–93	176 000
Somalia	Kenya	1993–94	110 000
Mozambique	Tanzania	1992–94	72 000
Angola	Zaïre	1991–92	60 000
Ethiopia	Kenya	1991–95	60 000
Togo	Benin	1994	50 000

Source: USCR, *World Refugee Survey* (produced annually) and other sources.

as a contributory factor in their resolve to force repatriation and restrict new arrivals, and it seems clear that some measure of force at least was used to ensure repatriation. However, in this case, there is plenty of evidence that security and geopolitical considerations were rather more important in determining the government's decision to press ahead with repatriation, rather than any particular concern with the environment (Pottier, 1998). Similarly, a United States Committee for Refugees report on Bangladesh notes how deforestation was cited by the Bangladeshi government as a reason for repatriation of the Rohingya refugees to Burma, along with the financial and staffing burden they placed on the state, increased crime, and the burden on local communities (USCR, 1995). But such pronouncements need not accurately reflect the 'truth' or context of a decision: in this case, the USCR points out how the $2.5 million that the Bangladeshi authorities claimed to have spent on refugee relief had in fact been 'compensated' by the UNHCR with a $3.2 million 'Affected Bangladeshi Villages Program' for non-refugee populations, whilst an extra 2000 government staff were in fact having their salaries paid by the UNHCR.

Based then on the starting point that large-scale return of refugees in the 1990s has been *largely* voluntary, and that where it has been forced, this is not clearly a result of environmental degradation, this chapter examines the experience of a number of these repatriations. These can be considered relatively orderly and gradual, compared to the experience of initial refugee flight, and this provides one element of interest from a natural resource management point of view. The next section considers two countries – Zimbabwe and Malawi – which were host to significant numbers of Mozambican refugees throughout the late 1980s to early 1990s. It asks whether in these countries, where much was written about environmental degradation associated with the refugees' presence, there has been a process of environmental rehabilitation, or recovery, once the 'population pressure' of these refugees was removed. Then, the chapter turns to the experience of Mozambique itself, as well as other case studies from Ethiopia, Bosnia and Guatemala, to examine the process of return and re-integration. In particular, the level of motivation of 'returnees' rather than 'refugees' to participate in natural resource management initiatives is considered, along with the broader economic, social and political context within which resource management takes place. Finally, the chapter tackles the question of how mass displacement – whether 'forced' at the start of persecution or conflict, or 'voluntary' at its end – relates to questions of natural resource management. Drawing on some of the arguments developed in previous chapters, the case of mass return helps to highlight the significance of social, economic or political 'dislocation' that accompanies refugee flight. It is suggested that there is a need to consider the way in which social institutions adjust to population movements of all kinds, rather than focusing solely on the creation, or subsequent actions, of refugees, in order to identify key influences on natural resource outcomes.

Environmental rehabilitation after refugees return

The return of refugees at the end of conflict signals, for host areas at least, the end of a period in which complex and arguably 'exceptional' arrangements need to be found to cope with additional populations' demands on the natural resource sector. In so far as refugees increase localised pressure of population on resources, their return can naturally be expected to release that pressure. Organisational and institutional measures established to cope with the presence of refugees are no longer required; host countries can turn to what many see as the more central issue of promoting local development. None the less, the transition from a refugee-hosting to a 'post-refugee' situation will not necessarily be straightforward. Most obviously, if damage has been caused to the environment or infrastructures of the hosting state, there is arguably a need to address that damage, and continue efforts to mitigate any negative impacts of the refugees' presence. However, more generally, the end of a refugee situation also implies the withdrawal of a range of international agencies, and what is often a significant investment in humanitarian assistance established to meet the immediate needs of the refugees. As such, the period of repatriation can also be a period of disinvestment in host areas by the international community. This can prove a testing time for international agencies themselves, as 'downsizing' their own operations has implications for cuts in staff and budgets. There are knock-on effects for national 'partner agencies', both in the government and non-governmental sectors. In short, no 'post-refugee' transition can be assumed to be smooth.

The exodus of refugees from Mozambique from the mid-1980s onwards was, at that time, one of the most significant mass displacements of population that the world had ever seen. As the country collapsed into war, violence and economic chaos, over one million refugees fled into neighbouring Malawi, whilst up to 250 000 found their way to Zimbabwe, of whom around 150 000 were settled in government-controlled camps. South Africa, Swaziland, Zambia and Tanzania also took significant influxes of refugees, with up to two million people displaced outside Mozambique, and three million inside the country itself (Wilson, 1992). These displacements were seen at the time as having major consequences for pressure on natural resources. As was noted in Chapter 4, in Zimbabwe, concern within the NGO community was sufficient for a new consortium of NGOs to be formed – the Fuelwood Crisis Consortium – to deal with issues of deforestation; whilst in Malawi, concern over pressure on woodfuel supplies had led to both the UNDP and the UNHCR funding large-scale programmes to supply woodfuel and to replant areas 'deforested' by refugees, respectively. At one stage, refugees represented up to one in ten of the total population in Malawi, in a country already widely seen as 'overpopulated' (M'manga, 1991; Kalipeni, 1992; World Bank, 1996b). At the same time, the refugee assistance operation in Malawi was a major foreign exchange earner for its government (Harrell-Bond, 1991).

The signing of the peace agreement between the two warring parties in Mozambique in 1992 signalled the start of a return to peace and stability in the region. By the time that democratic elections had occurred in 1994, a massive and voluntary repatriation of Mozambicans from Malawi and Zimbabwe (as well as other countries in the region) had begun. This in turn provided an opportunity for both countries to take stock of the impact of nearly ten years of hosting refugees, and in the natural resource sector, to plan for a future in which demand for woodfuel, timber and other forest resources would be dramatically reduced.[2] The process of transition to 'post-refugee' environmental management can be seen as complex in both countries. In Malawi in particular, it coincided with a period of democratisation and the end of single-party rule that had dominated the country since independence. One result has been increasing involvement of the international donor community in Malawi, with both positive and negative effects. External flows of aid and some private investment, a process of decentralisation of government and the rhetoric at least of participatory local development must be set against the rigours of structural adjustment and competition in world markets. Neither is irrelevant to the question of natural resource management at a local level, making a simple assessment of the effects of the refugees' departure impossible.

Analysis of environmental rehabilitation in both Malawi and Zimbabwe is facilitated though by the fact that in both cases, assessments were made, at the time of return, of the state of the natural environment in the former refugee-hosting areas. In Malawi, a large-scale study was commissioned by the EU, and conducted by the UK-based Natural Resources Institute (NRI, 1994). This report analysed both the causes of accelerated deforestation during the period in which refugees were present in Malawi and the specific situation faced by each of the major sites where the Mozambican refugees were settled. Whilst the scale of the 'impact' of refugees on local natural resources, highlighted in earlier reports (Wilson et al., 1989; Tamondong-Helin and Helin, 1991), was not disputed, the NRI report came up with a number of interesting observations. Thus in four districts where some of the largest concentrations of refugees were found during the war – Dedza, Mwanza, Chikwawa and Nsanje (Figure 7.1) – the NRI's conclusions were broadly that natural regeneration could be expected to 'repair' much of the damage that had occurred during the refugees' stay. In Dedza, host to around 175 000 refugees dispersed along the border with Mozambique, the report concluded that in spite of a negative effect on families displaced from their farm land by the presence of refugees,

> the area occupied by the settlements was insignificant in relation to the total area of cultivated land in the area concerned, and therefore the camps have had no impact on the degradation of arable land in general. The sites themselves were subject to erosion between the houses similar to other village areas. (NRI, 1994: 25)

In Dedza District, refugees had been settled on customary land that had previously been cultivated by local people, thus increasing the acute pressure on land resources. None the less, the NRI report concluded that no permanent damage had occurred, and that farming would restart on settled areas once the refugees returned.

In the other three main refugee-hosting districts, the story is much the same. In Chifunga, Luwani and Ndelema camps in Mwanza District (host to around 150 000 refugees), the report argued that 'no permanent damage has been done, and the sites will regenerate quickly if undisturbed' (NRI, 1994: 35), a scenario which was plausible given the isolation of the area, and the relatively low population density in contrast to Dedza. In Kunyinda camp in Chikwawa District, home to around 56 000 refugees, exactly the same conclusion was drawn; whilst in the densely populated Nsanje District at the southern tip of Malawi, the report suggested that:

> the only locally serious impact of the settlement areas has been to take up land which was previously used for grazing and to a lesser extent for cultivation. . . . Overall the impact has not been serious and when the refugees depart, the sites can be restored to their previous uses. (NRI, 1994: 30)

Such a conclusion was not particularly welcome to the Malawi government, which had hoped to lever international funds for the rehabilitation of areas affected by the refugees, and to continue the reforestation activities that had been carried out during the refugees' stay by the cash-starved Forestry Department. Moreover, there were some exceptions: for example,

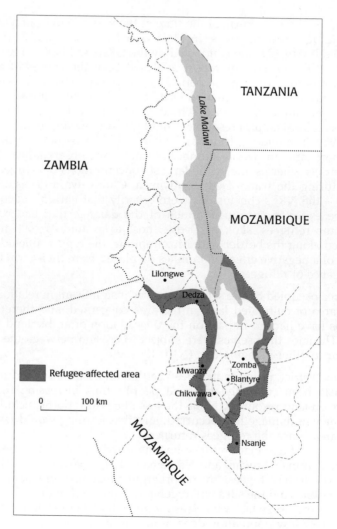

Figure 7.1 Major refugee-hosting districts in southern Malawi, 1984–94

in Chang'ambika camp in Chikwawa District, the NRI pointed out that increased exposure of the soil at a site already degraded before the refugees' arrival had added to the erosion rate. The report also raised the question of what use the land occupied by the refugees would be put to once the refugees had departed. In Mwanza, for example, the government had hoped to establish a rural 'growth pole', for resettlement of Malawians from other 'overpopulated' parts of the country. It could be argued that use of this relatively marginal area by refugees over a period of years had reduced its agro-economic potential, and that inputs of cash were

Figure 7.2   Abandoned refugee site in Mwanza District, Malawi
Source: (photo: Richard Black)

necessary to sustain subsequent settlement in the region. However, a more serious problem in this case appears to have been persuading Malawians that they wished to be resettled in Mwanza – in spite of the fact that facilities such as roads, schools, and a fully-equipped hospital at Luwani had been left after the refugees' departure (Figure 7.2).

In practice, the formal 'environmental rehabilitation' activities which took place in the refugee-affected areas of Malawi after the refugees' departure were largely limited to 'clean-up' operations at the individual sites occupied by the refugees. In Dedza District, some 24 703 refugee houses were demolished, pit latrines filled in, and sites then mechanically tilled to break the topsoil and make it suitable again for agricultural use (Le Breton et al., 1998a). Although some soil bunds were constructed to control soil erosion, in general there has been little attempt to take specific measures to restore fertility. This is perhaps not surprising, or of serious concern, given the manuring of the land over a period of time by both human and animal dung. Meanwhile, reforestation has often not been prioritised by local communities, not least since there remains quite a high degree of pressure on available agricultural land. None the less, in a participatory rural appraisal (PRA) of the area conducted by an organisation established by the local NGO community to oversee the rehabilitation process – the 'Coordination Unit for the Rehabilitation of the Environment' (CURE) – interest in tree planting and in loans to buy fertilisers and ox-carts was found (Mwale and Shaba, 1996).

Figure 7.3 Charcoal production in Mwanza District, Malawi
Source: (photo: Richard Black)

Elsewhere in Malawi, the same pattern of limited 'clean-up' operations, and subsequent use of the land according to local priorities and market conditions, appears to have been the rule. In general, this has implied some return to previous patterns of resource use, although this has not always been the case. In Mwanza, the pattern of change has been particularly interesting: here, although population densities dropped dramatically after the refugees' departure, deforestation has continued, largely as a result of charcoal production for the Blantyre market. This development reflects a major difficulty in making a simple pre- and post-refugee departure assessment of trends, as weakening restrictions on the trade in charcoal and a growing demand from urban centres have overlain any trends in resource use associated with the refugees' presence and subsequent departure (Figure 7.3). Indeed, a range of issues surrounding land tenure and access to natural resources remain in a state of flux, and perhaps deserve rather more direct analytical attention than ten years of the presence of refugees (Place and Otsuka, 1997).

A rather different pattern occurred in Zimbabwe, which owes much to the rather different nature of the refugee settlements there in comparison with Malawi. Whereas in Malawi, refugees were housed in a mixture of large, organised (but open) settlements and dispersed sites alongside Malawian villages, in Zimbabwe, the refugees were officially confined to five closed camps, each surrounded by barbed wire. Four of these camps were located in relatively isolated areas of communal land, and the other on the site of an abandoned agricultural research station.[3] They were run by a government ministry, the Department of Social Welfare, and although many refugees did leave the camps to search for work, especially in adjacent commercial farming areas, they were officially dependent on external support provided by the UNHCR.

As in Malawi, a study was conducted at the time of the refugees' departure, although in this case by the Fuelwood Crisis Consortium (FCC, 1994), the main actor involved in dealing with environmental issues during the refugees' stay. As in Malawi, this report again identified the potential for natural regeneration of the affected areas, although a number of specific rehabilitation activities were also highlighted. Overall, the FCC study concluded that some 12 000 ha of forest land had been converted to bush scrub savanna and scrub savanna. The study was based on a range of methods, including interpretation of air photographs taken in 1981, 1987 and 1992, overflying of the camps in 1994, consultation with local and central government departments and PRA exercises conducted in the refugee-affected districts, and in the case of Chambuta camp in southeastern Zimbabwe at least, it was not the first study to be based on such methods (Sparrow, 1993). However, despite firmer evidence of the extent of degradation around the Zimbabwean camps, and PRA exercises which did identify environmental rehabilitation priorities, actual 'rehabilitation' of the former refugee camps in Zimbabwe has been limited.

Once again, though, the lack of formal rehabilitation activities reflects less the release of population pressure from the areas concerned, and hence the absence of a perceived need to do anything about the sites, and more the development of new economic and political pressures in the period since the refugees' departure. Thus the Zimbabwean government, which had controlled the refugee camps much more tightly than in Malawi, has sought to put these to 'productive' use since 1994, in particular through the establishment of rural 'training centres' for diverse groups including street children, prostitutes and the elderly. In part, this reflected the institutional pressure noted above, in which the Department of Social Welfare needed a set of activities in order to avoid redundancies and a scaling down of its operations and importance. To date, though, the activities developed by the department, which has been transformed into a local NGO, have met with only limited success or funding. Problems have arisen as a result of the isolated location of the camps, and limited coordination between these planned uses (of the central administration) and the priorities identified in local development plans. Continued uncertainty about the future use of the sites has meant a lack of rehabilitation activities, or even of the removal of the wire fences that marked them out as camps. The only exception is the camp at Nyamatikiti in Rushinga District, where strong local lobbying has led to the transfer of the former refugee school to local community use.

In neither Malawi nor Zimbabwe does the notion of 'increased population leads to environmental degradation' seem to have been reversed as population numbers rapidly fell again with the repatriation of Mozambicans. Rather, signs of degradation in some areas continue to be in evidence (notably in Zimbabwe), whilst in other locations, a process of natural regeneration or productive use of the land for agriculture or other purposes has proceeded. Meanwhile, although during the refugee influx, natural resource management appears to have been problematic, there is little evidence that the refugees' departure has significantly improved the context for such management initiatives – or at least, this change has been overshadowed in Malawi in particular by the process of political change at a national level, and consequent upheaval in the policies and structure of local government and national resource management agencies mentioned above.

One significant development brought by the presence of refugees in both Malawi and Zimbabwe does deserve mentioning – namely the development of local NGO capacity in both countries to respond to natural resource management issues. In Zimbabwe, the Fuelwood Crisis Consortium, established to undertake measures to limit the environmental impact of the refugees during their stay, has subsequently evolved into an indigenous NGO (the 'Southern Alliance for Indigenous Resources', or SAFIRE), which has taken on an increasingly important role in developing local capacity for natural resource management. Similarly, in Malawi, CURE has also become a significant actor in this field, acting as a co-ordinating agency for other local and international NGOs working on natural resources, as well as providing training in PRA methods and other forms of capacity building for NGOs and a number of community-based organisations (CBOs). Without providing an in-depth assessment of their activities and impact, it is none the less interesting to note that both organisations owe their existence to the presence of refugees, and reflect a dynamic organisational response to forced migration that has had potentially longer-lasting consequences.[4]

Natural resource management in areas of return

Turning from the areas that refugees have left behind to the areas to which they return at the end of conflict and mass displacement, another interesting set of questions is raised about the capacity of areas of return to re-absorb large numbers of migrants, where the migrants concerned are no longer 'refugees'. In post-conflict situations, for example, there is less reason to believe that returnees should act as 'exceptional resource degraders' – since after all, they are returning to their own land. Moreover, the spirit of post-conflict reconstruction of war-damaged societies might be expected to *increase* motivation to rebuild institutions for environmental management, a task that may be particularly relevant given the tendency for both the 'environment' and social infrastructure to be damaged or destroyed during periods of war.

One important conceptual point is that in post-conflict countries, to which refugees return, the social, economic and political context for reintegration may well be just as dynamic as in the areas which refugees have left. An analysis of such situations needs to take into account issues as varied as the agro-ecological and socio-economic characteristics of areas of return, the broader administrative and legislative context, and the ways in which post-conflict countries become incorporated into, or excluded from, the global economy. The notion of population pressure on resources is no more appropriate in post-conflict countries than it is in any other part of the world; indeed, such areas are often the focus of renewed investment and incorporation into the world economic system, whilst the political factors which influence environmental outcomes may be complex. Where repatriation has been associated with the advent of democracy and free elections, it is likely that political aspirations will be high. Moreover, given the colonial and post-colonial history of marginal- isation of rural producers and land alienation in many countries affected by war in the 1980s and 1990s, such political aspirations are often centred around issues of land and other natural resources.

The return of Mozambicans from neighbouring countries in 1994 again provides an example of these processes, with refugees returning to a country where, despite much war damage, natural resources were per- ceived by many to be plentiful, and a new spirit of reconciliation was in evidence. In turn, the repatriation process as a whole was judged by the UNHCR to have been a success (Crisp et al., 1996). However, the notion of refugees returning to their 'own' land can be questioned in two import- ant ways. First, it is clear that in many cases, refugees were either unable to return to their area of origin or it was difficult to say exactly where that 'area of origin' might be. Throughout the war, large numbers of Mozambicans had lived for extended periods in internal 'refugee' camps or on resettlement schemes, whilst before the war, 'villagisation' policies had already forced many to leave their homes (UNECA, 1991). Coupled with uncertainty about the risks of a return to conflict, and the multiple moves made by many households in the last three decades, the choice for many of where to return to was a difficult one. Whilst such issues were mostly resolved after a period of time, the risk of damage to the environment around large and sometimes chaotic reception centres was real enough.

The problem of the high level of mobility that had occurred in Mozam- bique prior to and during the war was compounded by overlapping claims and confusion over rights to land and natural resources in the post-war period more generally. For example, a series of 'layers' of land rights, relating to historical occupation and lineage membership, the colo- nial period, post-independence nationalisations and subsequent popu- lation redistribution during the war have made the whole issue highly complex, creating numerous 'fuzzy entitlements' to land and other nat- ural resouces (Gruffydd-Jones, 1997). In addition, the allocation of land during and after the war to commercial companies as 'concessions' for agriculture, hunting or forestry has further undermined traditional rights to land, with Myers (1994) suggesting that by 1994, more than half the land area of Mozambique had been granted in concessions or sold to commercial enterprises. This has compounded a situation generated by the war itself, in which Renamo is believed to have systematically killed elephants and rhinos and to have exported the ivory and horns to pay for arms or to repay its external backers (McCallum, 1991; EIA, 1992). Also, although there are local variations, the creation of large numbers of new female-headed households has necessitated a renegotiation of tradi- tional rules of access in many areas, since women would traditionally gain access to land through their husband or through a male relative (Baden, 1997).

As a result of these pressures, the state of natural resource mangement in present-day Mozambique is far from rosy, even though a review by O'Keefe et al. (1991) towards the end of the war suggested that the country faced few 'environmental problems' *per se*. In the field of forestry, for example, localised pressures during the war, caused by the fact that the timber industry was restricted to over-exploiting the limited safe areas around towns and in the northern Cabo Delgado Province, have given way to wider problems associated with an upsurge in commercial logging. This process, linked to the expansion of South African timber companies, has been largely beyond the control of a state forestry service that is still developing a regulatory framework and capacity, and has limited financial resources. The perhaps inevitable dilemma here is between delaying investment whilst an appropriate capacity to regulate environmental impacts is established, and the need to reconstruct the Mozambican economy and maintain the confidence of foreign investors. But at least this dilemma points to the need to consider natural resource management in post-conflict situations in its wider political and economic context, and not simply in terms of local developments or the impacts of returnees.

At the local level, one relatively successful example of the establishment of a basis for natural resource management and access and control over local resources is provided by the Magoé District in the north-western part of Tete Province. Here, returning refugees and local people who remained during the war have participated in a community-based project to conserve wildlife resources. The '*Tchumo Tchato*' (literally: 'our wealth') project, instigated by the Dirreção Nacional da Floresta e Faune Bravia (DNFFB), has received funding from the IUCN and the Ford Foundation, and aims to empower the local community to manage its local resources sustainably and profitably, using the CAMPFIRE model from neighbouring Zimbabwe (Le Breton et al., 1998b). The project arose from international concern at excessive hunting of game in the area, but also from local resentment against what were seen as heavy-handed anti-poaching measures implemented by a private game hunting operation which held a concession covering much of the district. In two years of the project's operation, illegal hunting by locals has been largely brought under control, and a share of the profits of the hunting concession has been shared with local communities through a local natural resources council. The project has become something of a flagship for the Ford Foundation in Mozambique, although its success has not at present been replicated elsewhere.

Analysis of the experience of *Tchumo Tchato* is premature after only a year or two of operation, but even during this relatively short period, some interesting caveats are revealed about the potential for involvement of returnees in community-based natural resource management projects more generally. For example, in a visit to the project in August 1997, it proved impossible to locate more than a handful of people – mainly elderly women – who would admit to having been refugees in Zimbabwe or Zambia at all, in spite of the fact that flight from the area during the war clearly took place. In part, this may reflect an unwillingness of able-bodied men to admit that they did not fight in the war. But in addition, the district also appears to have attracted more recent immigration from neighbouring parts of Mozambique, such that the definition of the 'local community' might itself be questioned. In such a situation, it is understandable if those living in the area now are keen to disassociate themselves from any suggestion that they have ever lived elsewhere for an extended period of time, since this might undermine their claim to be part of a project which is redistributing benefits of natural resource management to local communities.

More challenging still are situations elsewhere, where the immediate benefits of sustainable natural resource management in post-conflict situations are less apparent. For example, in the case of Ethiopia, environmental degradation has been a concern of governments for decades. Land reform by the Derg government after 1975 was partly orientated towards addressing land insecurity and a highly unequal distribution of land, and after the 1984–85 famine, an Environmental Reclamation Programme based on food-for-work provided by the World Food Programme and other donors, concentrated in the highlands of northern Ethiopia, became the second largest such programme in Africa (Ståhl, 1990: 5). However, such initiatives were criticised by small farmers as being 'top-down' and relying too heavily on government-controlled and imposed institutions and their leaders, and external evaluations were muted at best (Yeraswork, 1988). Since the fall of the Derg government in 1991, there has been a new emphasis on 'participation' of local people in environmental programmes and in the development of a National Conservation Strategy (NCS) established with the assistance of the International Union for the Conservation of Nature (IUCN, 1990). But despite a conscious effort to incorporate peasant knowledge, it is not clear that this has fully succeeded, with many small farmers remaining sceptical of the value of environmental rehabilitation initiatives (Hoben, 1996).

Against this policy context, the period since 1991 has witnessed the return to Ethiopia of significant numbers of refugees who had fled to neighbouring countries during the Mengistu era. Major responses to this return movement have included the provision of funding by the UNHCR for the South Eastern Rangelands Project (SERP) in the Kebri Dehar region of south-eastern Ethiopia, a $670 000 project which sought to increase capacity of absorption of pastoralists, measured in terms of increasing livestock productivity, food security, and enhancement of sustainable use of natural resources in areas of reintegration (Asrat, 1996). This project, as well as a series of 'quick impact projects' (QIPs), many of which had environmental components, aimed to attract returnees from transit centres to their villages of origin. However, such projects, initiated by the UNHCR and other international agencies dealing with the returnees, have been criticised for remained relatively orientated towards short-term goals (*cf.* Kirkby et al. (1997) on QIPs in neighbouring Somalia), and there are clearly problems in resolving the dilemma between, for example, the UNHCR's short-term mandate in return situations, its desire to act quickly and its limited funding, and the objective of establishing projects that are community-led and sustainable in the long term.

An exception to this gloomy picture is provided by the Tigray region of Ethiopia, where Kibreab et al. (1998) argue that a high level of community mobilisation generated through the environmental awareness activities of the Relief Society of Tigray (REST) has been crucial in ensuring acceptance and some durability of a series of environmental measures implemented in an area of return. Projects in Tigray include soil and water conservation, area closures for natural regeneration of woodland, and tree planting. In this case, the level of political mobilisation is high, partly as a result of the efforts of REST during the war and the famine of 1984–85, and its explicit linking of environmental rehabilitation with the task of regional and national reconstruction. However, Kibreab et al. warn that such

> natural resource management programmes cannot be sustainable unless steps are taken to prevent the factors that cause the problem of resource depletion from recurring. In Tigray it was not lack of consciousness about the environment that was the cause of resource depletion. It was rather due to limited access to adequate resources in the context of increasing population density and stagnant or rudimentary techniques of production. (Kibreab et al., 1998: 24)

In part, the cases of Mozambique and Ethiopia demonstrate the importance of community participation in design and implementation of natural resource management initiatives, and the need to examine environmental priorities within the context of wider livelihood issues for the populations concerned – concerns which mirror those already discussed in the context of refugee situations. In addition, though, they again raise questions about the definition of 'communities', and the broader social, economic and political context within which natural resource management takes place at both a local and a national level.

Meanwhile, whereas in the Magoé District of Mozambique, community difference appeared to be downplayed as local residents sought to participate in the shared management of natural resources, in other return situations, fragmentation of communities is evident along a variety of lines, both related to the war itself, and as a result of factors quite disconnected from the war. For example, the repatriation of refugees and the very limited return of the internally displaced in Bosnia-Herzegovina since the Dayton Peace Accord in 1994 has been characterised by continuing community tensions, not only on the more obvious grounds of 'ethnicity', but also on grounds of class, gender, precise area of origin, and whether or not individuals fled abroad during the war (Black et al., 1997). As a result, attempts by international agencies to target physical rehabilitation (in this case particularly of houses destroyed in the war) at returnees have provoked a number of difficulties that have risked exacerbating, rather than easing, intra- and inter-community tensions. It is also clear that the return process itself has been used by nationalist forces within the different ethnic communities to assist the process of staking claims to areas of territory where the original inhabitants have fled as a result of 'ethnic cleansing'. Turning to a more specifically 'environmental' issue in the sense of natural resource management in Bosnia, it also appears that the wave of house reconstruction, through creating considerable demand for timber, has encouraged those currently controlling forest resources (widely reputed to represent the 'mafia' within each of the Bosnian state entities) to cut excessive amounts of trees in order to tap into this market. The result has been a significant drop in the price of timber, and an as-yet unquantified loss of forests in Bosnia (as well as neighbouring Croatia).

The case of return from Mexico to Guatemala also highlights the issue of control over land rights and rights more generally of access to natural resources at the time of refugee repatriation, and how this can be a far from simple process. In this case, land rights have proved a key issue both for returning refugees and for the government, and has led to some more problematic outcomes in terms of natural resource conservation. For example, Lawson (1995) describes how returning refugees in Guatemala refer to the 'repopulation' of the north of the country, rather than their 'reintegration', since they have been unable to return to their own land or villages. Although the Guatemalan government has offered returning refugees access to land for more than a decade (Manz, 1988), this has been associated with the clearance of 'new' areas of rain forest and the establishment of external control over areas rich in tropical hardwoods and a range of mineral resources.

Analysis of environmental change after conflict

It is not only the UNHCR that has developed a concern with post-conflict environmental rehabilitation initiatives in recent years. Other multilateral and bilateral development agencies have increasingly turned their attention to the aftermath of war and conflict as an important context within which development needs to be promoted. The World Bank has recently completed a number of studies into post-conflict repatriation and rehabilitation in Africa, and has developed new lending operations specifically to assist post-conflict reintegration of displaced people (Colleta et al., 1997). The International Labour Organisation has recently sponsored a major review of post-conflict employment opportunities, involving a focus on mainstreaming gender issues into employment promotion and skills training (Date-Bah, 1996). In reports on Bosnia, Guatemala, Lebanon and Mozambique, small-scale agriculture features as a significant sector in each case for rehabilitation and reconstruction activities (Baden, 1997; Loughna and Vicente, 1997; Nauphal, 1997; Walsh, 1997). The Development Assistance Committee of the Organisation for Economic Co-operation and Development has recently developed guidelines on conflict, peace and development co-operation which include environmental rehabilitation as a 'key programme component' in post-conflict situations (OECD, 1997: 59). Meanwhile, at the end of the 1980s, the UNRISD initiated a programme of research to investigate socio-economic aspects of rehabilitation in Africa (Allen and Morsink, 1994). The list is potentially endless, and given the significance of natural resources in the livelihoods of rural people in post-conflict situations, all of these initiatives have pointed directly or indirectly to the need to take 'environmental' reconstruction into account; although within the general concern with 'reconstruction' and 'rehabilitation', the natural resource sector has perhaps remained something of a side issue.

In the midst of such policy attention, and examples of practical reinteg-ration and reconstruction noted above, the question remains of how best to analyse and theorise issues relating to environmental change in condi-tions of post-conflict population return. The 'environmental entitlements' framework (Leach et al., 1997a), which links with emerging theory in 'new institutional economics', provides one useful basis on which analysis of post-conflict countries can be extended to address such environmental management issues. The framework uses an extended form of entitlements analysis, based on the pioneering work of Sen (1981) on livelihoods and entitlements to food during famines, to explore the ways in which differ-ently positioned social actors command environmental goods and ser-vices. It aims to enhance the current poor understanding of the dynamics of institutional arrangements in determining environmental outcomes. At the same time, theoretical advances of new institutional economics (NIE) provide a basis for understanding how organisational arrangements, rules and patterns of behaviour, both formal and informal, affect access to and use of natural resources. The argument which follows builds upon the examples cited in this chapter and the discussion in Chapter 6 of the importance of local-level institutions in natural resource management in the refugee situations of Guinea and Senegal, as well as a stimulating review provided by Gruffydd Jones (1997) on rehabilitation of rural live-lihoods in post-conflict contexts.

Management of natural resources and the explanation of accelerated environmental change (including environmental degradation) are cen-tral to the 'environmental entitlements' framework, which is based on the premise that the stock of available resources in a region is not syn-onymous with the use that is made of those resources. Sen's original formulation of the 'entitlements' approach highlighted how rural people and households have a range of 'endowments' – land, livestock, labour, knowledge, skills – but how it is important to focus also on the range of 'entitlements' that they can derive from the use of these endowments. Four categories of entitlement exist according to Sen, through exchange, production, labour and inheritance or transfer, such that any individual or household's 'entitlement set' is dependent not only on the endowments that exist but also on a process of 'entitlement mapping' which 'defines the possibilities that would be open . . . corresponding to each ownership situation' (Sen, 1981: 3). In the case of famines, a failure of entitlement (e.g. lack of ability to exchange, produce, work for or receive food) is what defines famine, rather than any particular absence of food itself.

When applied to environmental change, such an approach is helpful in that it highlights how any individual or household's use (or abuse) of natural resources will depend not simply on the presence/absence or ownership of resources, as is commonly assumed, but on access to resources through a variety of forms of 'entitlement', including exchange, production, labour and inheritance. Thus issues such as the influence of an individual's or household's position in society on their access to natural resources, and the rules which render claims over natural resources legitim-ate, can be seen as having crucial importance for subsequent management of resources. For Leach et al. (1997a), this implies a focus on a range of institutions, both formal and informal, which regulate natural resource management. Institutions can be defined as a range of 'regularised patterns of behaviour between individuals and groups in society' (Mearns, 1996: 106), and as such include religious and cultural norms, patterns of participation in market structures, and the development of intra-household patterns of resource use and structures of power between men and women.

It is important to point out that a focus on 'institutions' is not the same as a focus on 'organisations', even though the two terms are often used interchangeably. Indeed, there is perhaps an inevitable tendency in academic writing and especially amongst policy-makers, but also reflected in the discussion in this book, to overemphasise the role of 'organisations' and 'policy' at the expense of less visible, but arguably more important, 'institutions'. Such institutions often connect into the daily lives of rural people in a way that more formal organisations rarely achieve, and to which many would not even aspire. It is in this sense that the force of Mearns' argument can be seen, that 'a change in the configuration of institutions can make all the difference in net social product as well as the distribution of benefits among different resource users' (Mearns, 1996: 106).

Such an approach to 'environmental entitlements' contains some key differences to Sen's original formulation, which are brought out by discussion in a specifically environmental context. However, they are refinements that have also partially been raised within the context of direct responses to Sen's explanation of famine. For example, Devereux (1996) explores how in many African situations, there is a 'fuzziness' of entitlements both in terms of property rights and in terms of units of analysis, that were insufficiently dealt with by Sen. Indeed, in many societies across the developing world, there are often complex and overlapping rights to use, manage and dispose of different natural resources, which do not necessarily correspond to a Western or legalistic notion of 'property'. Equally, to focus on the household, whether in explaining famine or environmental degradation, risks ignoring the actions of, or consequences for, particular household members; and given existing patriarchal and hierarchical structures present in many societies, one result is often the marginalisation of women and/or children. Thus Devereux (1996: 4) draws a distinction between 'eligibility rules' which determine which individuals, households or communities have rights even in principle to use natural resources; 'queuing rules', which determine how long an individual may have to wait to use those resources, based for example on age or gender; and ability to pay – an increasingly important criterion that may even override the previous two. In each case, institutions can be seen as the key route through which access to resources is obtained.

In addition, to coin a phrase currently popular in a Western world rediscovering 'communitarian' philosophies, 'rights' to use resources often carry with them 'responsibilities' – including responsibilities for managing resources in a way that they are able to be transmitted to future generations. Perhaps more so than for rights, responsibilities can be seen to be governed not simply by law but by an array of formal and informal institutions which evolve over time. Rights and responsibilities towards the environment can be held by individuals, households, communities, or sections of communities. But whilst their combination often owes much to principles of reciprocity, governed by a desire to hedge risks and establish fall-back positions in times of uncertainty, the dynamic process of change in such institutions also often implies some measure of conflict between different actors. In this sense, institutional change in patterns of natural resource management can be seen as a dialectical process in which conflicts are both generated and resolved.

If such an approach is useful in promoting understanding of environmental outcomes in 'normal' situations in rural parts of the developing world, there are various ways in which it can be seen as particularly relevant not only to what are seen as 'post-conflict' situations but also to analysis of environmental issues during armed conflict and mass displacement as well. For example, in focusing on the role of institutions, the concept of transaction costs (the costs of obtaining information, searching out economic opportunities, and placing trust in other parties) is identified as central by work within 'new institutional economics' upon which the 'environmental entitlements' framework draws. Yet transaction costs are almost inevitably higher in situations of risk and uncertainty (such as during and after conflict). In this context it is interesting to examine – both theoretically and practically – whether the rational response is to retreat from formal transactions and rely more on social arrangements, or whether the transaction costs implied within informal social institutions are themselves put under intolerable strain during and after conflict situations.

In addition, what is perhaps most interesting is how the displacement engendered by armed conflict situations (and the end of armed conflicts) provides an additional dimension to the community fragmentation noted above, by potentially alienating 'refugees' from their 'hosts', or in the case of post-conflict return, 'stayers' from 'returnees'. In both cases, the risk is high that policy interventions, by prioritising assistance to those who are displaced, whilst ignoring those who host them or remain, will exacerbate these community divisions. Meanwhile, conflict itself, by breaking up existing networks of social organisation, drafting men and boys into armies, and challenging the authority of traditional figures such as chiefs and elders, can contribute to this process. However, if the flexibility and adaptability of natural resource management institutions are intimately related to, or even depend on, conflict between different social actors, it could be argued that refugee situations and subsequent mass return are far from representing some sort of 'special case' to be isolated from studies of environment and development. Rather, they can be seen as absolutely central to the environmental debate.

Applying the 'environmental entitlements' approach

The field of analysis of 'environmental entitlements' is very much in its infancy, and to date has rarely been applied specifically to situations of mass forced displacement or post-conflict repatriation. One collection based on research in progress at the Institute of Development Studies (IDS) at the University of Sussex focuses on cases studies in India, Ghana, South Africa and Zimbabwe (Leach et al., 1997b), two of which could be regarded as 'post-conflict', although neither deals specifically with areas of refugee return. As a group, these studies suggest that by focusing on the differentiation within local communities, it is possible to identify formal and informal mechanisms through which resources conflicts are resolved or mitigated. This can provide a basis for the promotion of more appropriate 'community-based' initiatives, which do not assume the shared interest of communities.

In one of the IDS studies on Ghana by Afikorah-Danquah (1997), there is a focus on immigration to the study area as a salient factor in community differentiation. In this case, migrants from the Sahel region to the north, who might be described by some as 'environmental refugees', are seen as contributing to environmental degradation through their involvement in charcoal burning and heavy short-term exploitation of the land. In a parallel to some of the arguments in Chapter 1 about 'exceptional resource degraders', it is suggested that the immigrants' lack of legal ownership of resources contributes to their destructive activities. However, the example of Guinea provided in Chapter 6 of this volume stresses how arrangements between local farmers and refugees can lead to 'environmental entitlements' for strangers that are not necessarily destructive, and indeed which strengthen local institutional capacity to manage natural resources. Even in Afikorah-Danquah's study, it appears that charcoal burning by immigrants is as much a product of contracts between local owners and immigrant specialists in charcoal production as it is in the latter's lack of secure resource ownership.

Similarly, although the studies by Cousins (1997) and Matose (1997) on South Africa and Zimbabwe, respectively, highlight the significance of resource ownership and citizens' control over resources, such a conclusion is not the only one to be drawn from the IDS collection. For example, Ahluwalia (1997) stresses how in India, local political mobilisation can act both across and within traditional caste, gender and class differences, whilst Kepe's (1997) analysis of conservation in the Eastern Cape of South Africa highlights the need for a trade-off between the different – and equally legitimate – conservation goals of different sections of the community. The point is that refugee migration, and post-conflict return, are not processes that automatically complicate natural resource management, either as a result of tenurial insecurity, high population density, or by virtue of the suddenness or unpredictability with which they occur. Rather, both are issues to be taken into account in any wider analysis of local or international conflict or co-operation over the use of the environment. The practical basis for this, and further implications for 'theory' in the field of ecology and environmental management, are developed further in the final chapter.

---

1. As with refugee flows in Chapter 1, 'large-scale' is defined here as involving flows of more than 50 000 people.

2.   The discussion of Malawi, Zimbabwe and Mozambique which follows is based on publically available documents, as well as informal discussions and interviews carried out during a visit by the author to the two countries in June–August 1997. The visit formed part of a broader project entitled 'Towards Sustainable Environmental Management Practices in Refugee-Affected Areas', carried out by the UNHCR. The views expressed in this chapter are those of the author, and not those of the UNHCR.

3.   A sixth camp at Chambuta, in Chiredzi District, was planned and developed, but never occupied due to the return of the refugees.

4.   A review of the experience of development of local NGO capacity during the refugee influxes in Malawi and Zimbabwe is provided by Harrell-Bond and Zetter (1993), who noted that beyond the specifically environmental field, a range of local NGOs and consortia were established at this time. However, they also warned that these NGOs generally continued to rely substantially on foreign donors, trapping them in an unequal power relationship and limiting their ability to genuinely alter the pattern of NGO activity.

CHAPTER 8

Linking research and practice

The central task of this book – to examine the complex interrelation-ships between refugee migration, environmental change and sustainable development – might be seen by some as an unusual or perhaps even an impossible task. Conventional wisdom has it that war, conflict and the production of refugees are (or at least should be) aberrations in the normal course of events in the developing world. Where the 'rule of law' (the rules governing what is considered 'civilised society') does break down, the response might be seen as relatively straightforward, at least in a conceptual sense. Conflict needs to be resolved, preferably through an integrated approach that considers the root causes of war and forced migration. Where these root causes include environmental degradation, or a development path that is leading towards environmental crisis, meas-ures are required to reduce environmental damage; this might involve technical measures, or the promotion of more equitable access to, and benefits from, natural resources. Meanwhile, in the countries or regions that become host to significant refugee populations, mitigative measures are also required to reduce environmental damage, increase the efficiency of use of natural resources, and establish a technical and organisational capacity on the part of external agencies to help refugees and local people prevent the further propagation of environment-related conflict. Such tasks might be seen to have little to do with the priorities of sustainable develop-ment, as exceptional circumstances require exceptional remedies. Whilst perhaps learning from experience of dealing with environmental issues elsewhere, policy is often necessarily developed 'on the spot' to deal with particular situations as they arise.

But that is not the message of this book. It is not intended to underes-timate the problems faced by refugees, host communities, governments or international agencies in responding to mass population displacement; or to pretend that easy solutions can be found that obviate the need, on occasion, to develop policies in an *ad hoc* and responsive manner. Clearly the arrival of 170 000 Rwandans at one border crossing point in Tanzania in one day, described in Chapter 5, would probably have placed a severe strain on whatever administration, national or international, was present to deal with it. None the less, a central argument developed in previous chapters is that humanitarian emergencies are *not* so exceptional, and that discussion of them can take place in the same context and applying some of the same principles as in the wider field of sustainable development. This has important implications for policy choices, since at least in principle, it suggests that the range of policy options available for natural resource management in general can be brought to bear on refugee situations. There is also a theoretical implication, as discussed in the previous chapter. In short, there is a case to be made that dynamic change in refugee-affected areas can tell us much about the workings of natural resource management systems and environmental change in a broader context.

The question of whether forced displacement is seen as exceptional or not is a key one then both for policy towards refugees (of which the main focus here, environmental policy, is just one part) and for theoretical approaches towards sustainable development. Many of the issues raised in this book have pointed towards the need for a more 'developmental' approach to environmental change in refugee-affected areas. The notions of the 'relief to development continuum' and a 'developmental' approach to refugee assistance in general are hardly new, having their origins in discussions surrounding the ICARA II conference, and, amongst others, the work of Tristan Betts (1984). However, more recent is the adoption of more 'developmental' language into the discourse of refugee policy. Thus whilst 'participation' is now as much a buzz-word in the UNHCR as it is elsewhere in the UN system, words such as 'stakeholder', 'co-management' and 'empowerment' still have some way to go in being seen as appropriate to refugee assistance. In turn, there has been relatively little enthusiasm or indeed interest to date in applying the notion of 'sustainable development' to refugee situations – except in the sense that 'sustainable livelihoods' are seen as what refugees and other displaced people might aspire to in post-conflict situations.

In this chapter, an attempt is made to draw together the concepts of humanitarian relief and sustainable development, in part by linking two other important and often separated fields of concern, research and practice. In the next section, the question is addressed as to why various policy options that are open in situations of forced displacement have not been built upon to develop an approach based on 'sustainability' or 'sustainable development'. The intention is not to repeat the discussion of specific policies presented in Chapter 4, but to build on this discussion, and the analysis of case studies in Chapters 5–7, to address the key strengths and weaknesses of, and obstacles to, an alternative approach to refugee policy that prioritises 'sustainable development'. Then, attention is turned to the way in which population–environment–development linkages are examined more generally in academic discourse, and the extent to which these concepts are generally applicable in refugee situations. The point of linkage for each of these fields can be seen to be less the answers to specific questions raised so far (how many environmental refugees are there; or what are the environmental impacts of refugees, and appropriate mitigative measures?) but rather the type of question raised in the first place. This issue, the scope of academic and practical enquiry and action, is addressed in the final section.

Practical concerns and policy options

As noted above, the notion of a 'developmental' approach to refugee assistance is not new. However, it is also worth noting that it has hardly become accepted practice. It could be argued that far from relief agencies becoming more developmental in their outlook, development agencies have become more and more concerned with relief. In part this reflects the rising number of refugees in the world through much of the 1980s and early 1990s, and a Western public and donor government response to media portrayal of 'disasters' in developing countries. Thus the share of humanitarian relief in the total aid spent by the UK, the USA and other OECD countries rose until 1994/95, mirroring the rise in the number of refugees, although both have subsequently declined slightly (Michel, 1996). This shift to humanitarianism also needs to be placed in the context of what some have described as the 'death' of development as a global discourse capable of mobilising world, or better, affluent northern opinion (Escobar, 1995; Gardner and Lewis, 1996). Arguably what has replaced it in a post-modern world is the more limited goal of alleviating immediate suffering associated with war, famine and disease.

In this sense, to repeat the call for a 'developmental' approach to environmental issues in refugee-affected areas (in other words, to call for a focus on sustainable development) might be seen at best as naïve, and at worst as ignoring the mistakes and misconceptions associated with the failure of such an approach to take hold after at least 10, and perhaps 20 years of trying. For example, McSpadden (1998) notes that the Eritrean government's insistence on refusing relief aid for repatriating refugees from the Sudan unless it was linked to national development priorities has effectively resulted in a lack of external funding. Although the rights and wrongs of this situation are debatable – and indeed hotly debated in this case – the decision to refuse to accept external funding is arguably much harder in a refugee context than in the context of return. Few are the countries, or the assistance agencies, who would turn down international assistance for refugees in order to stick to the principle that such assistance must contribute to a wider sustainable development goal.

The question is important as to why such a developmental approach has proven difficult to achieve, and why in particular it has proven unattractive to donors. For example, one of the consequences of the shift from development to humanitarian aid is that a number of more mainstream development agencies have been forced as a result to become more involved in humanitarian assistance. It might be expected that such agencies would support the integration of 'development' criteria, and a process of learning and exchange between the two sectors. For Zetter (1995: 50), the problem lies partly in the mandates of relevant agencies, and particularly what is now the lead agency for humanitarian assistance: 'as long as the operational remit for UNHCR, and thus its implementing partners, remains as an emergency relief agency, then shelter and settlement provision will always highlight the contradiction between the long and the short term interests'. One of the key long-term concerns, according to Zetter, is the environment, because of a combination of factors characteristic of refugee camps: 'high population densities, large population concentration and settlement on land which has often been neglected in the past precisely because of its inherent fragility'.

The mandates of international agencies are clearly important, yet such a position does not explain why the UNHCR in particular has been willing to take on environmental concerns, or why the rhetoric of 'environment', noted in Chapter 1, has not been translated into a rhetoric (or reality) of 'sustainable development'. Nor does it explain why development agencies' involvement in refugee situations has not spearheaded a change of attitudes towards the 'relief to development' continuum. Of course, a cynical view might hold that donor governments (and indeed the Western public) have no real interest in solving humanitarian crises and promoting development. As some measure of such 'development' has been achieved, for example, in south-east Asia, the result has been widely perceived as a competitive threat that underpins elements of decline in Western economies themselves – notwithstanding recent economic difficulties in that particular region. However, such a view is perhaps too cynical of the beliefs and intentions that motivate 'aid' in the first place; whilst some explanation at least for the failure of mainstream agencies to 'developmentalise' humanitarian relief can be found in much simpler operational realities. Thus in their eagerness to obtain a toehold in humanitarianism, to protect their workers' jobs, expand operations and be seen to be responding to public concern, it is easy to understand how development agencies might see it as necessary to accept the 'rules of the game' of the sector they are entering, rather than 'impose' their own views based on activities that most international observers still view as distinct and different. In this sense, it is the imagined 'mandate' of the sector in which an agency operates that can be seen as important, rather than necessarily the core mandate of the agency itself.

Of wider significance than mandates is the notion of what a 'developmental' approach might be capable of achieving, if it were adopted. It is easy, for example, from within a 'humanitarian' perspective, to consider its failings, and then to argue that more attention should be paid to 'successes' elsewhere in the development field. But this ignores the limitations within the 'development' field itself. For example, why is it that development aid in general is falling? Why is it that some authors are talking of the 'death' of development – of the rise of some new era of 'globalisation'? Moreover, three key issues are highlighted in this book which are of practical relevance to refugee assistance programmes, and which draw in part from developmental approaches. Thus it has been argued that there is a need for flexibility and participation of concerned actors; detailed awareness of local environmental circumstances; and knowledge of historical context. Each of these principles has long been recognised as relevant to the provision of emergency assistance to refugees, yet each has often been sidelined in the face of operational realities.

In some respects, it is a matter of judgement as to whether such sidelining was inevitable in any given situation, given the pressure under which international and national agencies implement their programmes. However, a number of more general points can be made as to why each of these principles is at best problematic.

Taking first the question of flexibility of programmes, and in particular flexibility and capacity to respond to the expressed needs of stakeholders through participatory analysis, this has become a key element of development assistance in recent years, and has permeated at least in principle into refugee assistance discourse. However, the rhetoric of 'participation' has not gone unchallenged, both as a philosophical principle and as a methodological guide. From a donor perspective, Eyben and Ladbury (1995) identify four reasons why there is often little real participation: first, it is often not in the economic interest of populations to participate; second, the social and political coherence of local 'communities' is often overestimated; third, participation can challenge the professional status and knowledge of agency staff promoting it; whilst fourth, people may be more interested in participating in some things (e.g. management of a productive resource) than others (e.g. management of a social service). It is all very well to promote 'participation', but too often in the development field this represents 'incorporation' of rural people into pre-defined project goals. As Lane (1995) points out, 'participation' is very much a creature of northern NGOs, widely accepted in donor circles but sometimes even resisted at local level. Meanwhile, for most of its advocates, 'participation' of communities involves a lengthy process, for which there is perceived to be little time in refugee situations.

Allowing space for local environmental and historical specificity in project development has also proven a highly problematic notion for the development field. The golden age of development thinking was very much one of globalising solutions that could be applied anywhere: import substitution industries, export-led growth and structural adjustment packages have given way to democratisation, decentralisation and institutional reform (i.e. privatisation) as the catch-all solutions that drive development forward. In this sense, refugee agencies have arguably been more successful than development agencies in finding place-specific solutions to problems, and a move towards a more 'developmental' approach in terms of what this has meant in practice might even be a retrograde rather than a positive step. What is required is not a 'blueprint' solution to the environment, which ignores local realities, but a set of policies that can take into account the diverse nature of ecosystems as much as the diverse (and often conflictual) needs and aspirations of people.

Whatever policy is promoted by international agencies, there is a danger that host governments will simply continue to react instinctively – and probably short-sightedly – to the presence of refugees. Where refugees can be used for cheap labour, they may be allowed to work, but not with the rights (and responsibilities) that might allow development of comprehensive and inclusive community-based systems of natural resource management. Part of the task therefore is to put to host governments some of the benefits that can be accrued, both from refugee agriculture, for example, and from refugees' involvement in the management of natural resources. At the same time, it is important to recognise that most natural resource 'management' probably goes on *despite* government or agency policy rather than because of it. Whilst formal 'community-based natural resource management' schemes such as Zimbabwe's CAMPFIRE programme may be difficult or impossible to set up in refugee situations, we should not be blind to the development of co-operative initiatives at a community level, which occur anyway. External policy could recognise and support these initiatives where they have beneficial environmental and refugee protection outcomes – although there is a need for caution too, since formal 'co-option' of such schemes into public policy may even destroy the reciprocal community base on which they survive. It is also important to recognise that 'community-based' does not necessarily mean 'good' either for the environment or for refugees, or for equity and development more generally. In each case, analysis is required, not the blind implementation of what appears to be the 'politically correct' solution.

Moreover, having regard for local realities means not only local environmental conditions but also the political and social context within which management of the environment takes place. For example, even if environmental policy promoted by external agencies is based on a sound analysis of the ecological realities of a region, if they are implemented with disregard for local laws, rules or procedures, they are not only likely to be doomed from a management point of view but they are also ethically highly questionable. A case in point was the Ukwimi Refugee Settlement in eastern Zambia mentioned in Chapter 4, where in addition to a reforestation programme, an attempt was made by one international agency to protect a 'forest reserve' that was recognised by neither the government nor the local population (Black et al., 1990). The result was widespread resentment of the rules not only by refugees (against whom they were aimed) but also by locals (who found themselves incidentally affected), and a wider rejection of the agency's environmental judgement by many in both groups. More recently, when environmental task forces were established in the Dadaab camps in Kenya, and in the Kagera and Kigoma regions in Tanzania, questions have arisen as to what authority they have to enforce rules against tree cutting and other activities considered environmentally 'destructive'.

These cases raise some key questions in the development of environmental management strategies in refugee-affected areas, which for that matter are not dissimilar to difficulties faced more generally in encouraging the 'greening of development'. For example, an environmental 'working group' or 'task force' can be helpful in the sense of bringing together actors to discuss common issues of concern, but without a traditional or legal framework within which to operate, it arguably has no 'right' – whether on grounds of traditional or legal justice – to intervene directly to influence the behaviour of individuals, groups or organisations towards the environment or natural resources when this is not provided for within the law of the country concerned. And yet the working group or task force's membership may see a failure to intervene as undermining their very purpose in meeting, and the consequences of intervention as so unquestionably benign or beneficial as to justify their action, even if this is without formal legal foundation.

There are many potential responses to this dilemma. Perhaps most obviously, such working groups or task forces can turn their attention (at least initially) to areas where they do or can have clear jurisdiction. For example, they can seek to influence the behaviour of the agencies whose representatives participate in meetings, or who have formally committed themselves to adhere to the 'rulings' of the group at the time of its establishment. Perhaps the best strategy to achieve this is likely to be in the form of promoting 'best practice' of agencies rather than formally laying down rules, although this will clearly depend on the institutional context within which agencies are operating, and the strength of feeling and argument about the need to deal with agencies that reject or ignore such 'best practice'. But it is also possible for even formal organisational frameworks such as working groups or task forces to act as a vehicle to link policy-makers to the range of informal institutions that were argued in Chapter 7 to be of crucial importance in natural resource management. This implies a role as a 'learning organisation', rather than strictly as a tier of 'management'.

Ethical issues in promoting sustainability for refugees

In addition to some of the practical reasons noted above why implementing solutions based on the principles of 'sustainability' in refugee situations has been problematic, there are also a number of broader reasons why this has been the case. In particular, an approach that prioritises 'sustainability' or conserving the environment and natural resources must itself start with a number of premises that are not necessarily straightforward. For example, in Chapter 1, it was noted how, of the 'strong' and 'weak' sustainability approaches, orientated respectively towards more 'ecological' or more 'human welfare' related goals, it is generally the latter that can be seen as more appropriate in refugee situations. And yet it is remarkable how in the available literature, and indeed in much of the discussion of policy experience in this book, it is ecological and especially biomass issues that have taken centre stage. In spite of agreeing with the principle of prioritising the *livelihood* environment, the reality all too often seems to be one that seeks to conserve first the intrinsic values of natural resources. It is easy to fall into the 'trap' of caring for trees – or better still, rare and endangered species of plants and animals – rather than taking care to ensure adequate supplies of woodfuel or animal protein for future generations.

Of course, such a dilemma is not one that is unique to refugee situations. However, it is perhaps ethically more urgent in the refugee field, given the sometimes desperate nature of refugees' circumstances. For example, if it is accepted that biodiversity is one of the key global environmental issues (and the current weight of world opinion and the essence of documents such as Agenda 21 suggest that it is) there might still be a case to be made that refugee situations should be seen as the exception to the rule, in which the short-term interests of a particular group *should* be allowed to take priority over global communal interests. Even if this were accepted, measures could be taken to try to avoid unnecessary pressure on irreplaceable natural resources, and operational agencies could be sensitised to the issue. But at the crunch point, if refugees arrive in a place where they threaten a certain resource, and there is nowhere else for them to go, operational agencies may be forced to weigh up the value of human life in the short term against the cost of environmental damage in the long term. And it is at this point that 'environmentalist' and 'humanitarian' perspectives might be expected to collide.

One way in which humanitarian agencies such as the UNHCR have tried to face up to such issues is to seek advice from the field of environmental economics. The environmental economist's response to such a dilemma is likely to be relatively simple: both human life and the long-term importance of an environmental good or service can be valued, in monetary terms. If the 'precautionary principle' with respect to biodiversity is strictly applied, the outcome is inevitable: biodiversity is infinitely valued, and humans (i.e. refugees) lose out. But even if concrete values are calculated for the natural resource in question, at levels well below infinity (based, for example, on its potential tourist value in the case of wild animals), refugees are still the likely losers, since the almost inevitable effect of placing a monetary value on life is to discriminate against the poor, weak and powerless. As Harvey (1996) points out, money is a form of 'social power', and the poor's lack of it is why environmental ills – and especially pollution and toxic waste – are disproportionately borne by the poor. In a refugee situation, the opportunity cost of a refugee's forgone earnings is unlikely to be high, through exactly the same form of reasoning as Laurence Summers' now famous World Bank memo on why toxic waste should be exported to the Third World. Summers' reasoning was impeccable on economic grounds (Anon., 1992), but perhaps rather more questionable ethically. One alternative to this unhappy conclusion that refugees' lives are simply not as important as biodiversity would be to seek to make a cost–benefit calculation having ruled out the two most objectionable outcomes: destruction of either the resource or human life. But again, the almost inevitable outcome would be the displacement of the refugees – and ultimately the same question would arise: where could their settlement be economically justified by a suspicious and conservative host, unprepared to take risks?

This problem, of who is calling the shots on where refugees can settle, forms part of a broader ethical problem connected with attempts to promote an approach to refugee assistance more orientated towards 'sustainability'. This is the question of who is framing the problem analysis. Despite emphasis in this book, and in an increasing volume of published and unpublished literature within the refugee field, on the need for and desirability of 'participatory' approaches, the hard reality is that technocrats and 'experts' are leading the way on the environment. The complexity of scientific processes, and the demands of thinking both globally and locally, push towards an expert-led model of intervention, in which sustainability must be learnt, not known and applied by those most affected in a particular area. Even this would not be so bad were it not for the explosion of expert advice, and the apparently unco-ordinated way in which it is handed out. For example, in Tanzania, in addition to the six 'environmental impact assessments' and the documentary film mentioned in Chapter 5 that occurred at the time of the refugee emergency, there have subsequently been at least four follow-up assessments on environmental issues alone, conducted for the UNHCR, IFAD, Care International and USAID. Clearly some of these assessment teams simply did not know of the existence of the others. In Goma, Biswas et al. (1994: 1) suggest that theirs was the 'first serious attempt to assess the environmental impacts of refugees at the macro level', adding that 'we feel it is a great tribute to the UNDP leadership that a development agency promoting sustainable development was the very first in the world to identify this gap'. In fact, the team led by Biswas was in Goma only weeks prior to a separate mission led by Ketel (1994b) for the UNHCR, but neither report makes any cross-reference. This is not to imply any criticism of the individuals involved: another example is this author's own visit to Benaco in Tanzania in 1994, which went on in complete ignorance of another environmental study (Green, 1994) conducted at around the same time by a colleague from Sussex University!

Questions of scale

Another important practical issue underlying discussion in previous chapters, and which causes problems for any attempt to consider long-term sustainability issues in refugee situations, concerns the central geographical question of scale. For example, what exactly is the spatial limit of refugees' environmental impacts; or in the case of people moving away from environmental problems, how far do people need to go in order for them to be classified as 'migrants'. On the former question, concerning the supposed 'damage' caused by refugees, there is an argument that where damage exists, it is relatively localised. This is because its impacts are linked to the distances people can move on foot in a single day or overnight with whatever natural resources they are searching for (Hoerz, 1995a). Such an observation, if true, is important, both because it helps keep any environmental change that is associated with forced migration in perspective, and because it suggests that policy orientated towards, for example, protecting certain resources, need not involve large-scale secondary displacement of forced migrants in order to ensure that protection.

In another sense, though, the consequences, or potential consequences, of environmental change and competition for natural resources in areas of refugee settlements cannot be thought of solely in relation to the locally affected area. A case in point concerns the dispersal and/or concentration of refugees in host areas, and the broad point of how large a settlement is appropriate environmentally. Where refugees are placed in small but dispersed settlements, any 'impact' is likely to be spread out, which may cause difficulty for host governments. Thus Cosgrave (1996) argues against large camps partly on environmental health grounds, but also suggests that a main reason for smaller camps is so that refugees can rely on local resources rather than being dependent on inputs from outside. But such reliance on local resources implies the 'burden' of refugee assistance will be borne to a much greater extent by the host country. In contrast, large refugee camps transfer this 'burden' or 'impact' to the international community, even if this involves environmental costs – in terms of transport, or packaging of aid items – that may be much greater than the short-term impact on a local area.

At one level, the argument of this book is that dispersal of refugees across a wider area is not necessarily an inappropriate solution for governments of host countries or host communities. Despite a largely negative review of the experience of moving refugees in Africa to mostly relatively small, planned agricultural settlements (Kibreab, 1989), much more positive examples can be drawn, not least that of Guinea cited in Chapter 6, where over half a million refugees from one of the most violent wars in Africa were able to stay for seven years without any serious incidents or conflict between refugees and local people. Lest this be considered a peculiarity of West African 'hospitality', an example can also be drawn from the past in the Great Lakes, where Betts (1984: 10) notes of the 1960s wave of Burundian refugees that: 'initially settled in the malarial Ruzizi plain, they were later moved, through the agency of UNHCR and the UN Children's Emergency Fund (UNICEF) to forested uplands further north. There, fertile soils and the presence of an experienced local management agency ... contributed to achievement of a successful ... assistance effort'. Indeed, by 1966, according to Betts, refugees were paying taxes to the Zaïrean authorities.

However, even if the creation of large camps is felt necessary, it is not necessarily a foregone conclusion that these will be environmentally damaging or unsustainable. One alternative way of considering the large camps or settlements is to view them as medium-sized cities (Nimpuno, 1996). Admittedly these are cities that are often created very rapidly, and in which a city infrastructure (water, sanitation, food and fuel supply, etc.) must be quickly established. Yet it could be argued that provision of this kind of infrastructure is precisely the task that the UNHCR and the international community seeks to fulfil, in which case failures of aid delivery should be seen as just that, and not as inevitable consequences of the construction of a camp. In turn, the sustainability of a city depends on its economic base, and the hinterland with which it trades, according to classic geographical theory. In this sense, the sustainability – or conversely the environmental destructiveness – of a refugee camp might be better seen as dependent on the kind of economy that can be established in the new camp-city. In other words, sustainability is not dependent so much on the size of a settlement, but rather on its economic viability in terms of the activities which its inhabitants are able, or allowed, to carry out.

Turning to the question of environmental causes of migration, the choice in this book to concentrate on examples of international migrants is a somewhat arbitrary one academically. However, work by Turton (1988) on the Mursi of Ethiopia is an example of 'forced migration', and perhaps of 'environmental refugees' in the sense that underlying causes of displacement of these populations have included factors relating to environmental degradation. The main difference is that crossing an international border brings in issues of international law and policy associated with the forced removal of people from one state to another, and the centrality of a particular UN agency – the UNHCR – to their protection and assistance. This raises an important point: if sustainable development is to be promoted to avoid environmentally induced displacement, is this meant to apply only to long-distance, transboundary migrations, or are relatively local-scale displacements considered undesirable as well?

Theoretical concerns and research options

The practical and ethical questions raised so far in this chapter all concern the difficulty of responding to the supposed environmental consequences or causes of mass flight, and highlight some of the operational issues involved. And yet what also characterises the discussion above is a central starting point: the assumption that forced migration, and especially conditions of 'emergency', are inevitably linked to unsustainable environmental damage. Based on such an assumption, the practical task becomes one of avoiding or mitigating damage, so that either the 'burden' of hosting refugees is reduced, or environmentally induced migration does not happen in the first place. But this notion that environmental disruption necessarily causes migration, or is the consequence of mass forced displacement, is one that was questioned in Chapter 2 of this book. In part, this reflects a paucity of knowledge, for example of empirical evidence of environmental degradation in refugee-affected areas; but at the same time, it could also be argued that the lack of conclusive evidence one way or another is inevitable, given the kind of question asked of refugee–environment relations in the first place. If different questions were asked – rooted perhaps in environmental or social theory – the answers might be quite different.

For example, in addressing the relationships between forced migration and environmental change, one possible approach is to look for spatial and temporal correlations between periods of mass forced migration and environmental decline, and to infer from evidence of such a relationship that a causal link exists in one direction or the other. However, even if a statistical relationship could be established (a moot point, as discussed in Chapter 2), such an approach would be unsatisfactory in isolating these two variables from a host of other factors that influence both migration and environmental change. An alternative approach would be to start from the logic of a theoretical position, and seek answers to questions raised by theory. Two relevant theoretical positions were outlined in Chapter 1, namely those of a broadly 'Malthusian' or 'Boserupian' nature. Given that the key variable for each of these positions – an increase in population – is produced very rapidly at a local scale under conditions of forced migration, it seems reasonable to consider whether empirical evidence supports the claims of one or other position in terms of observed environmental outcomes.

However, although the evidence presented in this book arguably provides more support to a Boserupian world-view than to a neo-Malthusian position, the Boserupian approach also contains a number of fatal flaws. In particular, given the likely failure of any particular empirical situation to produce an outcome that is clearly either 'positive' or 'negative' for the environment and/or for productivity and sustainability of the ecosystem, there is a tendency to search instead for the obstacles or constraints that have prevented the 'ideal' theoretical outcome from occurring. Such a search for 'necessary conditions' for a particular theory to be applicable arguably encourages an oversimplification of reality in order to produce a coherent 'model' of what is happening. In this process, assumptions can remain unexplained, and exceptions ignored, as one or more favoured explanations draw attention. A case in point is the heavy emphasis currently placed by many who believe in the capacity of communities to respond to environmental or other stresses, on the significance of single issues such as property rights or 'participation'. If such issues are dealt with, so goes the argument, communities *can* respond to scarcity in the way predicted by the Boserupian model. But by closing our eyes to the openness of natural and human systems, or to the 'failures' of other areas of policy or social systems, we may overlook what are ultimately the most important factors.

Thus although starting from an interest in a Boserupian theory of the response of communities to population growth and pressure on resources, the argument in this book has sought in practice to move some way towards a different approach. The key point is to examine once again what is the initial research question of importance. If the question is 'does forced migration cause environmental change?' the only reasonable answer is 'sometimes yes, sometimes no', with no easy way to make generalisations about when either outcome will be the case. However, if the question is 'in what ways are environmental issues of importance in forced migration?' (or vice versa), then more precise, if temporally and spatially specific, answers can be found. For example, it has been argued that at an international level, and for a variety of political reasons, the two areas of concern (environment and refugees) have become important in turn to international agencies whose primary concern lies elsewhere. This has driven a new field of international activity and intervention that has not always had desirable consequences. At a local level, similar linkages have occurred, as access to natural resources has become a major factor in refugees' livelihoods, and the presence of refugees in a region has stimulated institutional change in natural resource management systems, which has broader political ramifications.

For example, in rejecting the notion of refugees as 'exceptional resource degraders' in Chapter 1, and calling into question the theoretical usefulness of concentrating on refugees or forced migrants as social actors distinct from other groups, the purpose is not to deny the salience of forced migration in influencing environmental change altogether. Nor is calling into question more simplistic accounts of environmental degradation leading to forced migration and 'environmental refugees' meant to imply that environmental factors are not important in stimulating or perpetuating flows of migrants from fragile or degraded places around the world. But what both of the above points do imply is that discussion of the relationship between forced migration and environmental change needs to deal at a more complex level than trying to isolate simple causes and effects. Most importantly, there is no 'control' population against which the impact of forced migrants on the environment can simply be measured, any more than there is a control area against which the impact of environmental change on propensity to migrate can be measured. Rather, both processes are highly contextual, and demand analysis in the light of broader social, economic and political processes affecting the region and societies being studied.

This is of key importance not only for refugee situations but also for our understanding of the way in which political and social action evolves more generally. For example, in discussing the importance of property rights or participation for natural resource management, it is important to look outside the immediate context of the relationship of one action, institution or structure to another. Just as van den Bremer et al. (1995: 109) observe for rural Senegal how 'local participation in forestry activities serves to maintain political relations which may eventually bring other, more interesting opportunities', so too discussion of 'participation' in refugee contexts needs to situate this process in a wider socio-political context. This process of contextualisation – the examination of complex, overlapping and sometimes contradictory trends at a range of spatial scales – is a feature of writing in what can be loosely described as 'political ecology', which is discussed in the next section.

Approaches in political ecology

The field of 'political ecology' hardly represents a prominent or particularly coherent theoretical position – although the absence of a 'grand narrative' might be seen as one of its most positive features by those concerned with the universalising assumptions of five decades of writing about 'development'. But there are a number of key features of the approach outlined in detail by Bryant and Bailey (1997) that suggest its relevance to refugee situations, given the argument put forward in previous chapters. Thus Bryant and Bailey argue for the need to focus on the political role of different actors in human–environment relations, placing politics at centre stage, and seeking to examine local level issues in theoretical and comparative perspective. This ties in with the argument of this book, that what is more important is the way in which different actors influence social and environmental outcomes, rather than a technocratic approach to mitigation of particular environmental problems. They also stress the desirability of place-based, 'bottom-up' responses to environmental conflict, a view that also has some resonance here. Bryant (1997) focuses on what he describes as the 'impasse' of Third World environmental studies, in which there has been too much focus on proximate rather than ultimate causes. A 'political ecology' approach is seen as overcoming this impasse.

None the less, two key elements of the 'political ecology' approach outlined by Bryant and Bailey, and on which they argue all 'political ecologists' agree, might appear slightly less relevant to conditions of forced displacement. These are an emphasis on how environmental problems are linked to the world-wide spread of capitalism, and as a result, the need for 'far reaching changes to local, regional and global political–economic processes' (Bryant and Bailey, 1997: 3). Whilst this book certainly argues for a quite radical reassessment of the way environmental problems are viewed by some humanitarian agencies, there is a danger that this explicitly radical approach falls into a trap that Bryant and Bailey themselves warn against, of lapsing into an over-deterministic structuralist discourse, in which capitalism is seen as the root of all evil. In refugee situations, there is an opportunity in contrast to explore the ways in which a variety of actors, even relatively weak ones at a local level, can influence social and environmental relations. Perhaps the most important point is not to mould 'political ecology' into an overarching theory – indeed perhaps not to assume the label or agenda of 'political ecology' at all – but to apply some of its key concepts, such as a concern for place, the interaction of actors, and the historical and political context to practical problems.

Interestingly, the notion of 'political ecology' has its origins in the pioneering work of Piers Blaikie (1985) on another topic of very practical concern, in which prior thinking had also been highly problem-orientated. In turn, Blaikie's call for soil erosion to be seen in its political–economic context has a particular resonance in situations of forced migration, where issues such as access to land, security of tenure, and the nature of political relationships between different resource users are often in a state of great flux. At the same time, research conducted by geographers working in the broad field of 'political ecology' (Blaikie and Brookfield, 1987; Black, 1990; Watts, 1991; Bryant, 1992) indicates the extent to which even under 'normal' conditions (i.e. in the absence of forced migration), such relationships are frequently unstable, as well as how forced migration itself is historically much more widespread than a contemporary focus on the rising number of 'refugees' might lead one to expect. Thus in discussing the underlying 'causes' of soil erosion in terms of political–economic factors, Blaikie (1985: 127) talks of the process of 'spatial marginalisation', citing the example of colonial Kenya; whilst Blaikie and Brookfield (1987) cite examples involving state-sponsored resettlement in Malaysia, and transmigration in Indonesia, as cases in which large-scale population displacements have been associated with progressive environmental degradation over substantial periods of time.

One example of the application of the 'political ecology' approach to conditions of forced migration is provided by the work of Bassett (1988) in relation to environmental change, migration and conflict in the Sahel zone of northern Côte d'Ivoire. Thus for Bassett, the 'ultimate causes' of conflict in this region lie in an interrelationship between Fulani immigration, savanna ecology and the Ivorian development model, which has included surplus appropriation by foreign agribusinesses and the state, and misguided livestock development policies. For Bassett, a simple argument of resource scarcity causing conflict and migration is inadequate, and a multi-level analysis is necessary to explore interconnections between environmental, social and political change. But such an approach is not limited to those adopting an explicitly 'political ecology' framework, even within the discipline of geography. Thus the recent work of Hyndman (1997: 149), whilst not concerned specifically with environmental issues, none the less also seeks to situate practical outcomes for Somali refugees in the camps of northern Kenya within a wider geopolitical framework. Specifically, Hyndman examines the role of the 'global economy in refugee relief activities' in influencing highly localised events.

At the same time, in accepting the need to add a 'political' suffix to the term 'ecology', because of the extensive interrelationships between political, economic, social and environmental changes, it is important to resist attempts either to use the environment for political ends – or indeed to make unjustified claims to politicians for essentially 'environmental' ends. For example, a recent book by Welford (1997) entitled *Hijacking Environmentalism* describes how the international business community has latched onto environmental awareness in order to promote its own economic interests. Similarly, Swift's (1993) use of the term 'global political ecology' seeks to explain environmental crises as fundamentally rooted in a crisis of government and the political economy, implying that it is political change that is needed to head off impending environmental collapse. In neither case is the nature of environmental change itself really problematised: rather an accepted environmental problem is bounced back to the political community for action – or alternatively seen as resolvable if only there is 'political will' from the business community.

As has been argued in this book, there is a real risk of environmentalists and humanitarians treading the same path, in drawing alarmist conclusions about the environmental causes or consequences of forced migration in order to promote action about their own areas of concern. What is problematic about a potential and unholy alliance of environmentalists and humanitarians (or worse, anti-immigrationists), is that it risks obscuring the very real significance of environmental issues for many of the world's poorer migrants, by simply assuming, or repeating uncritically, popular belief about their nature and causes. At the same time, by assigning primacy to environmental issues, it risks deflecting public policy towards narrow responses which are unlikely to resolve the deep and complex problems that are facing many parts of the world today. A classic example is provided by Jacobson's (1988) assertion that John Steinbeck's 'Okies' in *The Grapes of Wrath* were the first 'environmental refugees'. This argument leads her to suggest that although the Depression was a factor in forcing massive migration from the mid-West to California in the 1930s, more important were 'unsustainable farming practices' and a 'severely degraded environment'. This almost technocratic reinterpretation of such an intensely political book is ironic in the extreme: one almost feels that American banks were right to foreclose loans made to small farmers, so that environmental management could be allowed to improve.

Work that explicitly refers to 'political ecology' is not the only theoretical field in which fruitful questions and lessons could be learnt about the relationships between refugees, environment and sustainable development. Another emerging field of theory within studies of environmental management and sustainable development concerns the significance of property rights in particular, and institutional regulation of access to natural resources in general, to environmental outcomes. Such concerns have their origin in the notion of the 'tragedy of the commons', in which Hardin's argument that 'freedom in the commons brings ruin for all' has been extraordinarily powerful (Hardin, 1968). Responses to Hardin's depiction of common property resources have been numerous, but some of the more interesting have focused on the capacity of management systems for such resources to control environmentally destructive practices effectively across a range of cultural and climatic contexts (Runge, 1984; Buck-Cox, 1985; Bromley and Cernea, 1989; Feeny et al., 1990). Following on from this though, there is a common perception that such management systems are increasingly under threat, or have already collapsed, in particular since they are based on the assumption of a 'closed system' with a clearly defined set of resource users largely isolated from broader economic and political influences. As areas are drawn into the world economy, see rapid population growth or, one might add, are subject to increasingly large population movements, the rules of reciprocity, respect for the common good and traditional cultural practice are seen as unable to cope with dramatic change. Even if refugees do not, at a particular moment in time, use resources in a more unsustainable way than autochthonous populations, their presence contributes to the breakdown of the structures that help to manage natural resource use.

However, this notion of the collapse of indigenous resource management systems – and of their inability to withstand change – also deserves to be held up to scrutiny. A number of recent studies have demonstrated how in places such as Machakos, Kenya (Tiffen et al., 1994), southern Africa (Scoones, 1996) and the forest–savanna transition zone in Guinea (Fairhead and Leach, 1996), traditional local systems of management have either survived or adapted and combined with a range of market mechanisms to ensure that environmental management goals have been supported. Given the dynamic socio-economic and political change stimulated by large-scale refugee movements, it is an interesting research question to address whether local resource management systems are capable of adapting to such change, and what might be the conditions of such adaptation. Such a focus on the dynamic process of institutional change in conditions of forced migration has also been one of the concerns of this book, and can start from the 'environmental entitlements' framework discussed in Chapter 7. In the West African case described in Chapter 6 at least, it seems reasonable to conclude that institutional adaptation is both possible and desirable from an environmental and social point of view.

Conclusion

This book, and this chapter, have sought to bring together practical and theoretical concerns in the examination of a specific problem – the relationship between forced migration, environmental change and sustainable development. Such a task is not straightforward, and there is perhaps an inevitable tendency to focus first on practical tasks, referring to theory when this is necessary to back up a view that emerges from 'empirical reality'. However, as noted in the introduction to this chapter, in many ways, it is this starting point, rather than necessarily the discussion that follows it, that is the key question in achieving a full understanding of the relationship. What matters more than evidence of environmental degradation 'caused' by refugees, or environmental 'causes' of forced migration, is an initial position that rejects a simple notion of a cause–effect relationship, and accepts the need to explore any social or environmental process in its local and global, and political as well as ecological context.

A number of interventions and policies have been explored here which seek to address refugee–environment relationships, and place such issues higher on the public agenda. However, without a critical view as to where environmental problems come from – what are their ultimate rather than proximate causes – there is a danger that environmental policy as it relates to forced migration simply becomes one more example of what Neumann and Schroeder (1995: 321) have described as 'a new sense of manifest destiny, a naturalized ecological mandate that drives environmental organizations and their donors to assert control over remote territories in new ways'. Rather than addressing complex relationships between refugee communities, their hosts, and host environments, the risk is that environmental policy becomes another way for states (or more likely, international organisations) to manage and control the often isolated areas in which refugees live. Whilst not denying that states need to exert some level of control over their borders, and over refugee populations that are accepted and protected within them, the establishment of strong environmental management can represent a precursor to, or work alongside, strong social management that ultimately threatens protection of refugees itself.

In seeking to move beyond this impasse, there are various 'simple solutions' that might be advanced to deal with refugee–environment relations, beyond the technical solutions reviewed in particular in Chapter 4. Perhaps the most obvious is that of promoting 'participation' of refugees and host populations in decisions concerning their local environments, and seeing the stock of natural resources in refugee-affected areas as primarily the 'livelihood environment' of these two groups. This is an approach that is explicitly favoured in this book, as an antidote to the belief in rational planning and control from the top down. But it is not an approach that is without problems. Thus Lipietz (1996: 219), in calling for a participatory and democratic approach to environmental problems (albeit primarily in an urban context), none the less highlights what is now well known and understood in Third World research: that 'those (men and women) below are not simply a homogeneous group ... but are themselves diverse and contradictory'. The point is that both social and environmental relations can also be viewed as power relations between different social groups. Nowhere perhaps is this more the case than in conditions of mass forced displacement.

However, despite the absence of any easy solutions, there remains a need to go further than simply identifying the difficulties of promoting environmentally equitable and socially just outcomes. There is a tendency, long noted, of critical (and especially radical) academic work to be predominantly 'oppositional' in the sense of finding flaws with development projects, actors, or the 'development' process itself (*cf.* Chambers, 1982). Yet criticisms of the application of 'development' or 'sustainable development' approaches too easily become an excuse for those who oppose any action to deal with poverty or inequality to do nothing at all. Similarly, an open mind on environmental questions can quickly be interpreted as justification for inaction, rather than for research to specify the nature of human–environmental interactions. The result is that practical action on environment and/or development issues by professional aid workers continues, largely uninformed, but still weakened by intellectual critique. Indeed, this is the only possible response to a position which questions the basic premises on which any kind of action takes place. In contrast, radical hopes of academics are pinned on 'community action'. And yet, at best, unrealistic hopes are often placed on such action, whilst at worst, the discourse of 'empowerment' itself becomes 'disempowering' (Fairhead, 1990) – especially when such action is isolated from sympathetic professional support. What is required instead is intellectual endeavour that engages with policy debates, to help set the terms on which public policy is carried out. Both the refugee and environmental fields provide important arenas in which such engagement can take place. For in the end, they represent areas where public support, though fragile, is strong enough not simply to critique the world but also, even if in small ways, to try to change it.

References

Abelson, P. (1996) *Project Appraisal and Valuation of the Environment: General Principles and Six Case Studies in Developing Countries*, Overseas Development Institute, London

Adams, A. (1977) *Le Long Voyage des Gens du Fleuve*, L'Harmattan, Paris

Adams, W. M. (1990) *Green Development: Environment and Sustainability in the Third World*, Routledge, London

Adisa, J. (1996) Rwandan refugees and environmental strategy in the Great Lakes region. A report on the Habitat/UNEP Plan of Action, *Journal of Refugee Studies*, **9**(3): 326–34

Afikorah-Danquah, S. (1997) Local resource management in the forest–savanna transition zone: the case of Wenchi District. Ghana. *IDS Bulletin*, **28**(4): 36–46

Ahluwalia, M. (1997) Representing communities: the case of a community-based watershed management project in Rajasthan, India. *IDS Bulletin*, **28**(4): 23–35

Alexander, J. and **McGregor, J.** (1996) *Our Sons Didn't Die for Animals. Attitudes to Wildlife and the Politics of Development: Campfire in Nkayi and Lupane Districts*, paper presented to International Conference on the Historical Dimensions of Democracy and Human Rights in Zimbabwe, 9–14 September 1996

Alexander, J. and **McGregor, J.** (1997) Modernity and ethnicity in a frontier society: understanding difference in northwestern Zimbabwe, *Journal of Southern African Studies*, **23**(2): 187–201

Allan, N. (1987) Impact of Afghan refugees on the vegetation resources of Pakistan's Hindukush-Himalaya, *Mountain Research and Development*, **7**(3): 200–4

Allen, T. and **Morsink, H.** (eds) (1994) *When Refugees Go Home*, James Currey, London

Anacleti, O. (1996) The regional response to the Rwandan emergency, *Journal of Refugee Studies*, **9**(3): 303–11

Anon. (1972) Limits to misconception: how seriously should we all take the forecast from the Club of Rome of impending world doom? in J. Blunden, P. Haggett, C. Hamnett and P. Sarre (eds) *Fundamentals of Human Geography: A Reader.* Harper & Row, London, pp. 84–6

Anon. (1992) Let them eat pollution, *The Economist*, 8th February 1992: 82

Ashley, C. (1992) War, logging and displacement in Burma (Myanmar), *Refugee Participation Network*, **12**: 24–7

Asrat, K. (1996) Refugee-related environmental projects in Ethiopia, in *Refugees and the Environment in Africa*, Proceedings of a workshop at Bahari Beach, Dar-es-Salaam, Tanzania, 2–5 July 1996, United Nations High Commissioner for Refugees, Geneva, pp. 22–3

Babu, S. C. (1995) International migration and environmental degradation: the case of Mozambican refugees and forest resources in Malawi, *Journal of Environmental Management*, **43**: 233–47

Bach, R. (1993) *Changing Relations: Newcomers and Established Residents in US Communities*, Ford Foundation, New York

Baden, S. (1997) *Post-Conflict Mozambique: Women's Special Situation, Population Issues and Gender Perspectives to be Integrated into Skills Training and Employment Promotion*, ILO Action Programme on Skills and Entrepreneurship Training for Countries Emerging from Armed Conflict, International Labour Organisation, Geneva

Barnes, D. F. (1994) *What Makes People Cook with Improved Biomass Stoves: A Comparative International Review of Stove Programs*, World Bank Technical Paper No. 242, World Bank, Washington, DC

Bassett, T. J. (1988) The political ecology of peasant–herder conflicts in the northern Ivory Coast, *Annals of the Association of American Geographers*, **78**(3): 453–72

Behnke, R. H., Scoones, I. and **Kerven, C.** (eds) (1993) *Range Ecology at Disequilibrium: New Models of Natural Variability and Pastoral Adaptation in African Savannas*, Overseas Development Institute, London

Benyasut, M. (1990) The *Ecology of Phanat Nihom Camp*, Indochinese Refugee Information Centre, Occasional Paper No. 1, Institute of Asian Studies, Chulalonghorn University, Bangkok

Betlem, J. (1988) *Le Changement de la Couverture Arborée des Forêts de Gonakier sur l'Ile a Morphil entre 1954 et 1986*, Notes Techniques No. 7, République Sénégal, MPN, DCSR, Pays-Bas, DGCI

Betts, T. (1969) Zonal rural development in Africa, *Journal of Modern African Studies*, **7**(1): 149–53

Betts, T. (1984) Evolution and promotion of the integrated rural development approach to refugee policy in Africa, *Africa Today*, **31**: 7–24

Binns, J. A. (1982) Agricultural change in Sierra Leone, *Geography*, **67**(2): 113–25

Biswas, A. K. and **Quiroz, C. T.** (1995) Rwandan refugees and the environment in Zaire, *Ecodecision*, Spring 1995: 73–5

Biswas, A. K., Tortajada-Quiroz, H. C., Lutete, V. and **Lemba, G.** (1994) *Environmental Impacts of the Rwandan Refugee Presence in North and South Kivu (Zaire)*, report submitted to the United Nations Development Programme (Office for Project Services), Regional Bureau for Africa, Zaire Country Office

Black, R. (1990) Regional political ecology in theory and practice: a case study from northern Portugal, *Transactions, Institute of British Geographers*, **15**(1): 35–47

Black, R. (1994a) Forced migration and environmental change: the impact of refugees on host environments, *Journal of Environmental Management*, **42**(4): 261–77

Black, R. (1994b) Environmental change in refugee-affected areas of the Third World: the role of policy and research, *Disasters*, **18**(2): 107–16

Black, R. (1994c) Refugee migration and local economic development: the case of Eastern Zambia, *Tijdschrift voor Economische en Sociale Geografie*, **85**(3): 249–62

Black, R. and **Mabwe, T.** (1992) Planning for refugees in Zambia: the settlement approach to food self-sufficiency, *Third World Planning Review*, **14**(1): 1–20

Black, R. and **Sessay, M. F.** (1995) *Refugees and Environmental Change: The Case of the Senegal River Valley*, Project CFCE Report No. 1, King's College London, Department of Geography, London

Black, R. and **Sessay, M. F.** (1997a) Refugees, land use, and environmental change in the Senegal River Valley, *GeoJournal*, **41**(1): 55–67

Black, R. and **Sessay, M. F.** (1997b) Forced migration, environmental change and woodfuel issues in the Senegal River Valley, *Environmental Conservation*, **24**(3): 251–60

Black, R. and **Sessay, M. F.** (1997c) From forest island to agricultural frontier? Forced migration and land-use change in the Forest Region of Guinea, *African Affairs*, **96**(385): 587–605

Black, R. and **Sessay, M. F.** (1998) Refugees and environmental change in West Africa: the role of institutions, *Journal of International Development*, **10**: in press

Black, R., Mabwe, T., Shumba, F. and **Wilson, K.** (1990) *Ukwimi Refugee Settlement: Livelihood and Settlement Planning*, unpublished report for Government of Zambia and NGOs, King's College London, October 1990

Black, R., Sessay, M. F. and **Milimouno, J. F.** (1996) *Refugees and Environmental Change: The Case of the Forest Region of Guinea*, Project CFCE Report No. 2, School of African and Asian Studies, University of Sussex, Brighton

Black, R., Koser, K. and **Walsh, M.** (1997) *Conditions for the Return of Displaced Persons from the European Union*, report to the European Commission, University of Sussex, Brighton

Blaikie, P. (1985) *The Political Economy of Soil Erosion*, Longman, London

Blaikie, P. and **Brookfield, H.** (1987) *Land Degradation and Society*, Methuen, London

Bloesch, U. (1995) Impact of humanitarian crisis on ecosystems (emphasis on vegetation), in *The Environmental Impact of Sudden Population Displacements: Priority Policy Issues and Humanitarian Aid*. Expert consultation sponsored by ECHO and organised by CRED, Université Catholique de Louvain, pp. 41–4

Bobb, F. S. (1988) *Historical Dictionary of Zaire*, Scarecrow Press, Metuchen, NJ

Borton, J., Brusset, E. and **Hallam, A.** (1996) *The International Response to Conflict and Genocide: Lessons from the Rwanda Experience. Study 3. Humanitarian Aid and its Effects*, Joint Evaluation of Emergency Assistance to Rwanda, Overseas Development Institute, London

Boserup, E. (1965) *The Conditions of Agricultural Growth: the Economics of Agrarian Change under Population Pressure*, Allen & Unwin, London

Bouchardy, J.-Y. (1995) *Development of a GIS System in UNHCR for Environmental, Emergency, Logistic and Planning Purposes*, Office of the Senior Coordinator on Environmental Affairs, United Nations High Commissioner for Refugees, Geneva

Bourque, J. D. and **Wilson, R.** (1990) *Guinea Forestry Biodiversity Study: Ziama and Diécké Reserves*, Ministère de l'Agriculture, République de Guinée, Conakry

Bouvier, A. (1992) Recent studies on protection of the environment in times of armed conflict, *International Review of the Red Cross*, **291**: 554–66

Bromley, D. and **Cernea, M.** (1989) The *Management of Common Property Natural Resources: Some Conceptual and Operational Fallacies*, World Bank, Discussion Paper 57, Washington, DC

Bruce, J. P., Lee, H. and **Haites, E. F.** (1996) *Climate Change 1995: Economic and Social Dimensions of Climate Change*, Contribution of Working Group III to the Second Assessment Report of the Intergovernmental Panel on Climate Change, Cambridge University Press, Cambridge

Bryant, R. L. (1992) Political ecology: an emerging research agenda in Third World studies, *Political Geography Quarterly*, **11**(1): 12–36

Bryant, R. L. (1997) Beyond the impasse: the power of political ecology in Third World environmental research, *Area*, **29**(1): 5–19

Bryant, R. L. and **Bailey, S.** (1997) *Third World Political Ecology*, Routledge, London

Buck-Cox, S. (1985) No tragedy on the commons, *Environmental Ethics*, **7**: 49–61

Burbridge, P. R., Norgaard, R. B. and **Hartshorn, G. S.** (1988) *Environmental Guidelines for Resettlement Projects in the Humid Tropics*, FAO Environment and Energy Paper No. 9, Food and Agriculture Organisation of the United Nations, Rome

Burkholder, B. T. and **Toole, M. J.** (1995) Evolution of complex disasters, *The Lancet*, **346**: 1012–15

Burnham, P. (1993) The Cultural Context of Rainforest Conservation in Cameroon, paper presented at the 36th Annual Meeting of the African Studies Association, Boston, Mass., 4–7 December 1993

Caldwell, L. K., Bartlett, R. U., Parker, D. E. and **Keys, D. L.** (1982) *A Study of Ways to Improve the Scientific Content and Methodology of Environmental Impact Assessment*, Advanced Studies in Scientific Techniques and Public Affairs, School of Environmental Affairs, Indiana University, Bloomington

Cambrézy, L. (1995) *Kakuma (Kenya): Preliminary Study for a Monograph of a Refugee Camp*, ORSTOM/UNHCR, Nairobi

Cannon, T. (1995) What makes emergencies different? Inter-relations of development, environment and disasters, in *The Environmental Impact of Sudden Population Displacements: Priority Policy Issues and Humanitarian Aid*, expert consultation sponsored by ECHO and organised by CRED, Université Catholique de Louvain, pp. 18–27

Cernea, M. (1990) Internal refugee flows and development-induced population displacement, *Journal of Refugee Studies*, **3**(4): 320–39

Chambers, R. (1982) *Rural Development: Putting the Last First*, Longman, London

Chambers, R. (1986) Hidden losers? The impact of rural refugees and refugee programs on poorer hosts, *International Migration Review*, **20**: 245–65

Chambers, R. (1997) *Whose Reality Counts? Putting the First Last*, Intermediate Technology Publications, London

Chisholm, C. (1996) *Refugee Settlements are an Economic Magnet to Host Populations: A Case Study of Meheba Refugee Settlement, North West Province, Zambia*, Student Field Trip Report, King's College London, Department of Geography, London

Colleta, N., Kostner, M. and **Widerhofer, I.** (1997) *Case Studies into War-to-Peace Transition: Ethiopia, Namibia and Uganda*, IBRD Discussion Paper No. 331, Washington, DC

Collins, A. E. (1996) *Environment, Health and Population Displacement in Mozambique: the Case of Cholera Bacillary Dysentery*, PhD thesis, King's College, London

Collins, R. O. (1990) *The Waters of the Nile: Hydropolitics and the Jonglei Canal, 1900–1988*, Clarendon, Oxford

Condé, J. and **Diagne, P.** (1986) *South–North International Migrations: A Case Study. Malian, Mauritanian and Senegalese Migrants from the Senegal River Valley to France*, Organisation for Economic Cooperation and Development, Development Centre Papers, Paris

Cordell, D. D., Gregory, J. W. and **Piché, V.** (1996) *Hoe and Wage: A Social History of a Circular Migration System in West Africa*, Westview, Boulder, Colo.

Cosgrave, J. (1996) Refugee density and dependence: practical implications of camp size, *Disasters*, **20**(3): 261–70

Cousins, B. (1997) How do rights become real? Formal and informal institutions in South Africa's land reform. *IDS Bulletin*, **28**(4): 59–68

Crewe, E. (1997) The silent traditions of developing cooks, in R. D. Grillo and R. L. Stirrat (eds) *Discourses of Development: Anthropological Perspectives*, Berg, Oxford, pp. 59–80

Crisp, J., Fahlen, M., Christie, F., O'Keefe, P., Danilowicz, J. and **de Wolf, S.** (1996) *Rebuilding a War-Torn Society: A Review of the UNHCR Reintegration Programme for Mozambican Refugees*, United Nations High Commissioner for Refugees, Geneva

Curran, P. J. and **Foody, G. M.** (1994) The use of remote sensing to characterise the regenerative states of tropical forests, in G. M. Foody and P. J. Curran (eds) *Environmental Remote Sensing from Global to Regional Scales*, Wiley, Chichester, pp. 44–83

Cutler, P. (1984) Famine forecasting: prices and peasant behaviour in northern Ethiopia, *Disasters*, **8**: 48–56

Daffe, M., Laura, P. and **Cisse, S.** (1991) *Étude de la Problématique du Bois Combustible dans le Département de Podor*, Rapport Final, SIC/SAS/ABF.MDRH, DEFCCS, DCCE, Dakar

Daley, P. (1991) Gender displacement and social reproduction: settling Burundian refugees in western Tanzania, *Journal of Refugee Studies*, **4**(3): 248–66

Date-Bah, E. (1996) *Sustainable Peace after War: Arguing the Need for Major Integration of Gender Perspectives in Post-Conflict Programming*, ILO Action Programme on Skills and Entrepreneurship Training for Countries Emerging from Armed Conflict, Working Paper, International Labour Organisation, Geneva

David, R. (1995) *Changing Places? Women, Resource Management and Migration in the Sahel, Case Studies from Senegal, Burkina Faso, Mali and Sudan.* SOS Sahel, London

Davidson, B. (1994) On Rwanda, *London Review of Books*, 18 August

Davies, S. (1996) *Adaptable Livelihoods: Coping with Food Insecurity in the Malian Sahel*, Macmillan, Basingstoke

Dbaké, C. (1980) *Introduction à l'Utilisation des Données d'Enquêtes et de Recensement dans l'Analyse des Migrations*, Direction de la Statistique du Sénégal, Dakar

De Graaf, J. (1997) *The Price of Soil Erosion: An Economic Evaluation of Soil Conservation and Watershed Development*, Tropical Resource Management Paper 14, University of Wageningen, The Netherlands

De Groot, W. T. (1989) *Participation and Environmental Management: An Exploration*, Studies in Environment and Development, Centre for Environmental Studies, Leiden University

Devereux, S. (1996) *Fuzzy Entitlements and Common Property Resources: Struggles over Rights to Communal Land in Namibia*, IDS Working Paper No. 44, Institute of Development Studies, Brighton

De Waal, A. (1997) *Famine Crimes: Politics and the Disaster Relief Industry in Africa*, International African Institute and James Currey, London

Dewees, P. A. (1989) The woodfuel crisis reconsidered: observations on the dynamics of abundance and scarcity, *World Development*, **17**(8): 1159–72

DHA (1995) *United Nations Consolidated Inter-Agency Appeal for Persons Affected by the Crisis in Rwanda. January–December 1995. Vol. II: The Sub-Regional Perspective*, Department for Humanitarian Affairs, Geneva

Diallo, M. S., Fischer, J. E., Koulibaly, A., Kourouma, I. K., Kourouma, S., Camara, Y., Lamah, P., Kpomy, A., Délamou, E. and **Camara, K.** (1995) *Le Foncier et la Gestion des Ressources Naturelles en Guinée Forestière: une Étude de Cas du Terroir de Nonah*, research report, Land Tenure Centre, University of Wisconsin, Madison

Diessenbacher, H. (1995) Explaining the genocide in Rwanda: how population growth and a shortage of land helped to bring about the massacres and the civil war, *Law and State*, **52**: 58–88

Diop, A. B. (1965) *Société Toucouleur et Migration*, Initiations et Études, No. 18, Institut Fondamental de l'Afrique Noire, Dakar

Döös, B. R. (1994) Why is environmental protection so slow? *Global Environmental Change*, **4**(3): 179–84

Dregne, H. E. and **Tucker, C. J.** (1988) Desert encroachment, *Desertification Control Bulletin*, **16**: 16–19

Dualeh, M. W. (1995) Do refugees belong in camps? *The Lancet*, **346**: 1369–70

Dufournaud, C. M., Quinn, J. T. and **Harrington, J. J.** (1994) A partial equilibrium analysis of the impact of introducing more efficient wood-burning stoves into households in the Sahelian region, *Environment and Planning A*, **26**: 407–14

Durrani, N. (1996) Income-generating project for refugee areas in Pakistan (IGPRA 1984–1996), in *Refugees and the Environment in Africa.* Proceedings of a Workshop at Bahari Beach, Dar-es-Salaam, Tanzania, 2–5 July 1996, United Nations High Commissioner for Refugees, Geneva, pp. 34–40

ECHO/CRED (1995) *The Environmental Impact of Sudden Population Displacements: Priority Policy Issues and Humanitarian Aid,* Expert consultation sponsored by ECHO and organised by CRED, Université Catholique de Louvain, Brussels

Eckholm, E. P. (1975) *The Other Energy Crisis,* World Watch Paper No. 1, World Watch Institute, Washington, DC

EIA (1992) *Under Fire: Elephants in the Front Line,* Environmental Investigation Agency, London

EIU (1996a) *Country Report: Zambia, Zaire,* 2nd Quarter 1996, Economist Intelligence Unit, London

EIU (1996b) *Country Report: Zambia, Zaire,* 3rd Quarter 1996, Economist Intelligence Unit, London

El-Hinnawi, E. (1985) *Environmental Refugees,* United Nations Environment Programme, Nairobi

ERM (1994) *Refugee Inflow into Ngara and Karagwe Districts, Kagera Region, Tanzania. Environmental Impact Assessment,* report by Environmental Resources Management, for CARE International and Overseas Development Administration, London

ERM (1995) *Environmental Interventions in Gikongoro Prefecture of Southwestern Rwanda, Final report, April 1995,* report by Environmental Resources Management, for CARE, London

Escobar, A. (1995) *Encountering Development: The Making and Unmaking of the Third World,* Princeton University Press, Princeton, NJ

Eyben, R. and **Ladbury, S.** (1995) Popular participation in aid-assisted projects: why more in theory than practice? in N. Nelson and S. Wright (eds) *Power and Participatory Development,* Intermediate Technology Publications, London, pp. 192–200

Fairhead, J. (1990) *Fields of Struggle: Towards a Social History of Farming Knowledge and Practice in a Bwisha Community, Kivu, Zaire,* PhD thesis, School of Oriental and African Studies, University of London

Fairhead, J. and **Leach, M.** (1994) Contested forests: modern conservation and historical land use in Guinea's Ziama reserve, *African Affairs,* **93**(373): 481–512

Fairhead, J. and **Leach, M.** (1995) False forest history, complicit social analysis: rethinking some West African environmental narratives, *World Development,* **23**(6): 1023–35

Fairhead, J. and **Leach, M.** (1996) *Misreading the African Landscape: Society and Ecology in a Forest–Savanna Mosaic*, Cambridge University Press, Cambridge

Falconer, J. (1990) *The Major Significance of 'Minor' Forest Products: The Local Use and Value of Forests in the West African Humid Forest Zone*, FAO Community Forestry Note No. 6, Food and Agriculture Organisation of the United Nations, Rome

FCC (1994) *An Assessment of the Environmental Damage in the Areas Surrounding the Mozambican Refugee Camps in Zimbabwe*, report prepared by the Fuelwood Crisis Consortium on behalf of the Agricultural Sub-Committee of the Mozambican Refugee Programme, Harare, Zimbabwe

Feeny, D., Berkes, F., McCay, B. and **Acheson, J.** (1990) The tragedy of the commons: twenty-two years later, *Human Ecology*, **18**(1): 1–19

Ferris, E. (1993) *Beyond Borders: Refugees, Migrants and Human Rights in the Post-Cold War Era*, WCC Publications, Geneva

Findlay, A. (1993) End of the Cold War: end of Afghan relief aid? in R. Black and V. Robinson (eds) *Geography and Refugees: Patterns and Processes of Change*, Belhaven, London, pp. 185–97

Findley, S. E. (1994) Does drought increase migration? A study of migration from rural Mali during the 1983–1985 drought, *International Migration Review*, **28**(3): 539–53

Foley, G., Moss, P. and **Timberlake, L.** (1984) *Stoves and Trees: How Much Wood Would a Wood Stove Save if a Wood Stove Could Save Wood?* Earthscan, London

Forbes Martin, S. (1991) *Refugee Women*, Zed Books, London and New Jersey

Formoli, T. A. (1995) Impacts of the Afghan–Soviet war on Afghanistan's environment, *Environmental Conservation*, **22**(1): 66–9

Foster, B. E. (1989) The Rwandese refugees in Uganda, in A. M. Ornas and M. A. M. Salih (eds) *Ecology and Politics: Environmental Stress and Security in Africa*, Scandinavian Institute for African Studies, Uppsala, pp. 145–55

French, D. (1986) Confronting an unsolvable problem: deforestation in Malawi, *World Development*, **14**(4): 531–40

Frischmuth, C. (1997) *Gender is Not a Sensitive Issue: Institutionalising a Gender-Orientated Participatory Approach in Siavonga, Zambia*, Gatekeeper Series No. 72, International Institute for Environment and Development, London

Gallagher, D. and **Forbes Martin, S.** (1992) *The Many Faces of the Somali Crisis: Humanitarian Issues in Somalia, Kenya and Ethiopia*, Refugee Policy Group, Washington, DC

Gardner, K. and **Lewis, D.** (1996) *Anthropology, Development and the Post-Modern Challenge,* Pluto, London

Germain, J. (1984) *Guinée: Peuples de la Forêt,* Académie des Sciences d'Outre-mer, Paris

Ghimire, K. (1994) Refugees and deforestation, *International Migration,* **32**(4): 561–70

Gill, J. (1987) Improving stoves in developing countries: a critique, *Energy Policy,* **15**(2): 135–43

Girard, S. (1992) Greening la Mosquita, *Refugees,* **89**: 18–19

Gitonga, S. (1995) Energy and refugee camps: options for interventions, *Boiling Point,* **37**: 3–6

Glassman, J. (1992) Counter-insurgency, ecocide and the production of refugees: warfare as a tool of modernisation, *Refuge,* **12**(1): 27–30

Glazovsky, N. F. and **Shestakov, A. S.** (1994) *Environmental Migrations caused by Desertification in Central Asia and Russia,* paper presented to Almeria Symposium on Migration and Desertification, United Nations Negotiating Committee for a Convention to Combat Desertification, February 1994, Almeria

GOM (1992) *Paper on the Impact of Refugees on Malawi's Environment, Social and Economic Infrastructure,* paper prepared by Government of Malawi for International Conference on 'First Country of Asylum and Development Aid', Blantyre, June 1992

Goma Epidemiology Group (1995) Public health impact of Rwandan refugee crisis: what happened in Goma, Zaire, in July, 1994? *The Lancet,* **345**: 339–44

Goodland, R., Mercier, J. R. and **Munteba, S.** (eds) (1995) *Environmental Assessment in Africa: A World Bank Commitment,* World Bank, Washington, DC

Grainger, A. (1993) *Controlling Tropical Deforestation,* Earthscan, London

Green, R. H. (1994) *That They May be Whole Again: Offsetting Refugee Influx Burdens on Ngara and Karagwe Districts,* report for UNICEF, September 1994, Ngara and Dar es Salaam

Green, R. H. (1995) *Making Whole Again: Human Development Consequences of Hosting Refugees in Kagera Region, Tanzania,* paper presented at ESRC Development Economics Study Group Annual Conference, University of Leicester, 24–25 March 1995

Grimsich, G. (1996) CARE Tanzania environmental programme: environmental projects in refugee-affected areas, in *Refugees and the Environment in Africa,* Proceedings of a workshop at Bahari Beach, Dar-es-Salaam, Tanzania, 2–5 July 1996, United Nations High Commissioner for Refugees, Geneva, pp. 49–55

Gruffydd Jones, B. (1997) *Post-Conflict Rehabilitation of Rural Livelihoods: Lessons from Africa*, MA Dissertation in Rural Development, University of Sussex, Brighton

GTZ/UNHCR (1992) *Domestic Energy and Reforestation in Refugee-Affected Areas: Sudan and Malawi, Main Report*, Mission Report, Deutsche Gesellschaft für Technische Zusammenarbeit GmbH and United Nations High Commissioner for Refugees, Geneva.

Guha-Sapir, D. and **Salih, M. A. M.** (1995) Environment and Sudden Population Displacement: Policy Issues for Humanitarian Action and Development Programmes, in *The Environmental Impact of Sudden Population Displacement: Priority Policy Issues and Humanitarian Aid*, expert consultation sponsored by ECHO and organised by CRED, Université Catholique de Louvain, pp. 7–17

Haack, B. and **English, R.** (1996) National land cover mapping by remote sensing, *World Development*, **24**(5): 845–56

Hanley, N. and **Spash, C. L.** (1993) *Cost–Benefit Analysis and the Environment*, Edward Elgar, Aldershot

Haque, C. E. and **Zaman, M. Q.** (1993) Human response to riverine hazards in Bangladesh: a proposal for sustainable floodplain development, *World Development*, **21**(1): 93–108

Hardin, G. (1968) The tragedy of the commons, *Science*, **162**: 1243–8

Harrell-Bond, B. E. (1986) *Imposing Aid: Emergency Assistance to Refugees*, Oxford University Press, Oxford

Harrell-Bond, B. E. (1991) A fresh approach, *Africa Events*, October 1991: 24–6

Harrell-Bond, B. E. and **Zetter, R.** (1993) *Aid, Non-Governmental Agencies, and Refugee Livelihood: Recommendations for a Way Forward*, a report for government and agencies on Mozambican refugees in Malawi and Zimbabwe, Refugee Studies Programme, Oxford

Harrell-Bond, B. E., Voutira, E. and **Leopold, M.** (1992) Counting the refugees: gifts, givers, patrons and clients, *Journal of Refugee Studies*, **5**(3/4): 205–25

Harvey, D. (1974) Population, resources and the ideology of science, *Economic Geography*, **50**: 256–78

Harvey, D. (1996) *Justice, Nature and the Geography of Difference*. Blackwell, Oxford

Hasler, R. (1993) *Political Ecologies of Scale and the Multi-Tiered Co-Management of Zimbabwean Wildlife Resources under Campfire*, CASS Occasional Paper Series, Centre for Applied Social Sciences, University of Zimbabwe, Harare

Hazarika, S. (1993) *Bangladesh and Assam: Land Pressures, Migration and Ethnic Conflict*, Occasional Paper of Project on Environmental Change and Acute Conflict, American Academy of Arts and Sciences, Washington, DC

HEAT GmbH (1994) *Technical Consultancy for the Introduction of Fuelwood Saving Stoves in Refugee and Orphan Camps in Goma, Zaire*, report prepared for GTZ, Eschborn

Helldén, U. (1991) Desertification: time for an assessment? *Ambio*, **20**(8): 372–83

Hoben, A. (1996) The cultural construction of environmental policy: paradigms and politics in Ethiopia, in M. Leach and R. Mearns (eds) *The Lie of the Land: Challenging Received Wisdom on the African Environment*, International African Institute and James Currey, London, pp. 186–208

Hoerz, T. (1995a) *Refugees and Host Environments: A Review of Current and Related Literature*, unpublished paper, Refugee Studies Programme, Oxford

Hoerz, T. (1995b) The environment of refugee camps. A challenge for refugees, local populations and agencies, *Refugee Participation Network*, **18**: 17–19

Hoerz, T. (1996a) *Kakuma Case Study – Debriefing Notes*, Mitigation of Environmental Degradation in Refugee Hosting Areas through Participatory Systems Research Project, GTZ, Nairobi, May 1996

Hoerz, T. (1996b) Enhancing participation in the management of camp environments, in *Refugees and the Environment in Africa, Proceedings of a Workshop Sponsored by the Office of the Senior Coordinator on Environmental Affairs*, United Nations High Commissioner for Refugees, Geneva, pp. 107–21

Hoerz, T. and **Kimani, M. J.** (1996) *Mitigating Environmental Degradation in Refugee-Affected Areas through Participatory Systems, A GTZ Research and Pilot Project*, Mid-term report, October 1996, GTZ, Nairobi

Hoerz, T., Kimani, M. J. and **Muthoni, W.** (1996) *Participatory Systems to Mitigate Environmental Degradation in Refugee Hosting Areas of East Africa*, report prepared for United Nations Environment Programme, Nairobi

Homer-Dixon, T. (1994) Environmental scarcities and violent conflict: evidence from cases, *International Security*, **19**(1): 5–40

Horowitz, M. (1991) Victims upstream and down, *Journal of Refugee Studies*, **4**: 164–81

Horowitz, M., Salem-Murdoch, M., Grimm, C., Kane, O., Lericoltais, A., Magistro, J., Niasse, M., Nuttall, C., Scudder, T. and **Sella, M.** (1990) *Suivi des activités du bassin du fleuve Sénégal, Phase 1, rapport definitif*, Institute for Development Anthropology, Binhampton, NY

Hubbell, D. and **Rajesh, N.** (1992) Not seeing the people for the forest: Thailand's program of reforestation by forced eviction, *Refuge*, **12**(1): 20–1

Hudson, H. (1992) Money trees in Pakistan, *Refugees*, **89**: 17

Hyndman, J. (1996) *Geographies of Displacement: Gender Culture and Power in UNHCR Refugee Camps, Kenya,* PhD thesis, Department of Geography, University of British Columbia, Vancouver

Hyndman, J. (1997) Border crossings, *Antipode*, **29**(2): 146–76

INCED (1994) *The Almeria Statement on Desertification and Migration, 11 February 1994,* Intergovernmental Negotiating Committee for a Convention to Combat Desertification, Châtelaine, Switzerland

INSEE-Coopération (1962) *Enquête Démographique au Senegal, 1961–62,* INSEE, Paris

IOM (1996) *Environmentally-Induced Population Displacements and Environmental Impacts Resulting from Mass Migration,* International Symposium, Geneva, 21–24 April 1996, International Organisation for Migration with United Nations High Commissioner for Refugees and Refugee Policy Group

IOM/RPG (1992) *Migration and the Environment,* International Organisation for Migration, Geneva and Refugee Policy Group, Washington, DC

Islam, M. (1992) Natural calamities and environmental refugees in Bangladesh, *Refuge*, **12**(1): 5–10

IUCN (1980) *World Conservation Strategy,* International Union for the Conservation of Nature and Natural Resources, United Nations Environment Programme and World Wildlife Fund, Geneva

IUCN (1990) *Ethiopia National Conservation Strategy, Phase I,* report prepared for the Government of the People's Democratic Republic of Ethiopia with the assistance of IUCN, Addis Ababa

IUCN (1991) *Protected Areas of the World: A Review of National Systems. Volume 3: Afrotropical,* International Union for the Conservation of Nature and Natural Resources, Venezuela

Jackson, C. (1993) Doing what comes naturally? Women and environment in development, *World Development*, **21**(12): 1947–63

Jacobs, M. (1991) *The Green Economy: Environment, Sustainable Development and the Politics of the Future,* Pluto, London

Jacobsen, K. (1994) *The Impact of Refugees on the Environment: A Review of the Evidence,* Refugee Policy Group, Washington, DC

Jacobsen, K. (1996) *The Role of the Host Government in Kenya and Uganda,* consultant's report, Mitigation of Environmental Degradation in Refugee Hosting Areas through Participatory Systems Research Project, GTZ, Boston, July 1996

Jacobsen, K. (1997) Refugees' environmental impact: the effect of patterns of settlement, *Journal of Refugee Studies*, **10**(1): 19–36

Jacobson, J. (1988) *Environmental Refugees: A Yardstick of Habitability*, World Watch Paper No. 86, World Watch Institute, Washington, DC

Jaspars, S. (1994) *The Rwandan Refugee Crisis in Tanzania: Initial Successes and Failures in Food Assistance*, ODI Relief and Rehabilitation Network Paper No. 6, Overseas Development Institute, London

Johnson, D. L. and **Lewis, L. A.** (1995) *Land Degradation: Creation and Destruction*. Blackwell, Oxford

Kakonge, J. O. (1994) Monitoring of Environmental Impact Assessments in Africa, *Environmental Impact Assessment Review*, **14**: 295–304

Kalipeni, E. (1992) Population growth and environmental degradation in Malawi, *Africa Insight*, **22**(4): 273–82

Kaplan, R. D. (1994) The coming anarchy, *The Atlantic Monthly*, February 1994: 44–76

Kelly, C. (1996) *Disasters and Environmental Change: The Impact of Population Displacement and Options for Mitigation*, paper presented to Pan-Pacific Hazards '96, Vancouver

Kennedy, W. V. (1988) Environmental impact assessment and bilateral development aid: an overview, in P. Wathern (ed.) *Environmental Impact Assessment: Theory and Practice*, Routledge, London, pp. 272–85

Kepe, T. (1997) Communities, entitlements and nature reserves: the case of the Wild Coast, South Africa. *IDS Bulletin*, **28**(4): 47–58

Ketel, H. (1994a) *Tanzania: Environmental Assessment Report of the Rwandese Refugee Camps and the Affected Local Communities in Kagera Region, 2–30 June 1994*, PTSS Mission Report 94/29N, United Nations High Commissioner for Refugees, Geneva

Ketel, H. (1994b) *Zaire: Environmental Assessment Report of the Rwandese Refugee Camps and Immediate Surroundings in North and South Kivu, 21 November–16 December 1994*, PTSS Mission Report 94/62/N, United Nations High Commissioner for Refugees, Geneva

Kharoufi, M. (1994) Forced migration in the Senegalese–Mauritanian conflict: consequences for the Senegal River valley, in S. Shami (ed.) *Population Displacement and Resettlement: Development and Conflict in the Middle East*, Center for Migration Studies, New York, pp. 140–55

Kibreab, G. (1989) Local settlements in Africa: a misconceived option? *Journal of Refugee Studies*, **2**(4): 468–90

Kibreab, G. (1994) Migration, environment and refugeehood, in B. Zaba and J. Clarke (eds) *Environment and Population Change*, International Union for the Scientific Study of Population, Derouaux Ordina Editions, Liège, Belgium, pp. 115–29

Kibreab, G. (1996) *People on the Edge in the Horn: Displacement, Land Use and the Environment in the Gedaref Region, Sudan*, James Currey, London

Kibreab, G. (1997) Environmental causes and impact of refugee movements: a critique of the current debate, *Disasters*, **21**(1): 20–38

Kibreab, G., Nauheimer, H., Walker, R., Jama, A. K., Medhanie, Z. and **Bruk, Y.** (1998) *Towards Sustainable Environmental Management Practices in Refugee-Affected Areas: Environmental Review in Southeastern and Northern Ethiopia*, draft report, Environment Unit, United Nations High Commissioner for Refugees, Geneva

Kikula, I. S. and **Magabe, A. B.** (1996) *Environmental, Social and Policy Implications of Refugees in Kagera*, unpublished paper, National Environment Management Council, Dar-es-Salaam

Kimani, M. J. (1994) *Household Energy and Institutional Energy Conservation Programme. Incorporating Development Concepts in 'Relief' Operations: A 'Keg of Powder' that Requires Only a 'Spark'*, final report, Energy Conservation Training Programme in the Goma Refugee Complex, Kahindo Refugee Camp, December 1994

Kirkby, J., Kliest, T., Frerks, G., Flikkema, W. and **O'Keefe, P.** (1997) UNHCR's cross-border operation in Somalia: the value of Quick Impact Projects for refugee resettlement, *Journal of Refugee Studies*, **10**(2): 181–98

Kunze, D., von Bergen, M., Blauꟙez, L., Haslwimmer, M., Hinterberger, J., Schaefer, S. and **Schmüdderich, C.** (1990) *Différenciation de la Population-Cible du Projet Kabare à la Base d'une Analyse Socio-Économique dans la Région du Kivu, Zaïre*, Seminar für Landwirtschaftliche Entwicklung, Technische Univerisität Berlin

Lachaal, L., Chowdhury, Q. and **Rouchiche, S.** (1995) *South Khorosan Rangeland Rehabilitation and Refugee Income-Generating Project, Joint UNHCR-IFAD Final Evaluation Mission, May 6–June 16, 1995*, Report to United Nations High Commissioner for Refugees, Geneva, International Fund for Agricultural Development, Rome, and the Islamic Republic of Iran, Tehran

Lacoste, Y. (1973) An illustration of geographical warfare: bombing of the dikes on the Red River, North Vietnam, *Antipode*, **5**(2): 1–13

Lamont-Gregory, E. (1995a) The environment, cooking fuel and UN resolution 46/182, *Refugee Participation Network*, **18**: 14–16

Lamont-Gregory, E. (1995b) Cooking fuel and the Rwanda crisis, *The Lancet*, **344**: 546

Lamont-Gregory, E., Henry, C. J. K. and **Ryan, T. J.** (1995) Evidence-based humanitarian relief interventions, *The Lancet*, **346**: 312

Lamotte, M. (1983) The undermining of Mont Nimba, *Ambio*, **12**: 175–9

Lane, J. (1995) Non-governmental organisations and participatory development: the concept in theory versus the concept in practice, in N. Nelson and S. Wright (eds) *Power and Participatory Development*, Intermediate Technology Publications, London, pp. 181–91

Lassailly-Jacob, V. (1993) Refugee–host interactions: a field report from the Ukwimi Mozambican Refugee Settlement, Zambia, *Refuge*, **13**(6): 24–6

Lassailly-Jacob, V. (1994a) Government-sponsored agricultural schemes for involuntary migrants in Africa: some key obstacles to their economic viability, in H. Adelman and J. Sorenson (eds) *African Refugees: Development Aid and Repatriation*, Westview, Boulder, Colo., pp. 209–26

Lassailly-Jacob, V. (1994b) *Scheme-Settled Refugees and Agro-Ecological Impact on Hosts' Environment, A Field Report from Ukwimi, an Agricultural Settlement in Zambia*, unpublished report, Centre for Refugee Studies, York University, Toronto

Lawas, C. M. and **Luning, H. A.** (1996) Farmers' knowledge and GIS, *Indigenous Knowledge and Development Monitor*, **4**(1): 8–11

Lawson, G. M. (1995) *Flowers from the Ash: The Communities of Population in Resistance and the Process of Reintegration in the Ixcán Jungle of Guatemala*, MSc thesis, University of Texas at Austin

Lazarus, D. (1991) Climatic change creates environmental refugees, *IDOC International*, **2/91**: 1–2

Leach, M. (1992) *Dealing with Displacement: Refugee–Host Relations, Food and Forest Resources in Sierra Leonean Mende Communities during the Liberian Influx, 1990–91*, IDS Research Reports No. 22, Institute of Development Studies, Brighton

Leach, G. and **Mearns, R.** (1988) *Beyond the Woodfuel Crisis: People, Land and Trees in Africa*, Earthscan, London

Leach, M. and **Mearns, R.** (1991) *Poverty and Environment in Developing Countries: an Overview Study*, Final Report to Economic and Social Research Council, Swindon

Leach, M. and **Mearns, R.** (eds) (1996) *The Lie of the Land: Challenging Received Wisdom on the African Environment*, International African Institute and James Currey, London

Leach, M., Mearns, R. and **Scoones, I.** (1997a) Environmental Entitlements: A Framework for Understanding the Institutional Dynamics of Environmental Change, IDS Discussion Paper No. 359, Institute of Development Studies, Brighton

Leach, M., Mearns, R. and **Scoones, I.** (eds) (1997b) Community-based sustainable development: consensus or conflict? *IDS Bulletin*, **28**(4), Institute of Development Studies, Brighton

Le Breton, G. (1992) Carrying a heavy load, *Refugees*, **89**: 8–11

Le Breton, G. (1995) Stoves, trees and refugees: the Fuelwood Crisis Consortium in Zimbabwe, *Refugee Participation Network*, **18**: 9–12

Le Breton, G., Black, R., Nabane, N. and **Stevenson, K.** (1998a) *Towards Sustainable Environmental Management Practices in Refugee-Affected Areas*, Country Report for Malawi, draft report, Environment Unit, United Nations High Commissioner for Refugees, Geneva

Le Breton, G., Black, R. and **Nabane, N.** (1998b) *Towards Sustainable Environmental Management Practices in Refugee-Affected Areas. Country Report for Mozambique*, draft report, Environment Unit, United Nations High Commissioner for Refugees, Geneva

Lemarchand, R. (1994) Managing transition: Rwanda, Burundi, and South Africa in comparative perspective, *Journal of Modern African Studies*, **32**(4): 581–604

Leopold, L. B. et al. (1971) *A Procedure for Evaluating Environmental Impact*, United States Geological Survey Circular No. 645, US Department of the Interior, Washington, DC

Lericollais, A. (1989) Risques anciens, risques nouveaux en agriculture paysanne dans la vallée du Sénégal, in M. Eldin and P. Milleville (eds) *Le Risque en Agriculture*, Orstom, Paris

Lericollais, A. and **Vernière, M.** (1975) L'emigration toucouleur: du fleuve Sénégal à Dakar, *Cahiers Orstom Série Sciences Humaines*, **12**(2): 161–75

Leusch, M. and **Burie, A.** (1996) Refugees and the environment in North-Kivu, Zaire, in *Refugees and the Environment in Africa*, Proceedings of a workshop at Bahari Beach, Dar-es-Salaam, Tanzania, 2–5 July 1996, United Nations High Commissioner for Refugees, Geneva, pp. 59–62

Lipietz, A. (1996) Geography, ecology, democracy, *Antipode*, **28**(3): 219–28

Lo, C. P. (1986) *Applied Remote Sensing*, Longman, London

Long, L., Cecsarini, L. and **Martin, J.** (1990) *The Local Impact of Mozambican Refugees in Malawi*. Report to USAID and the United States Embassy, Lilongwe

Loughna, S. and **Vicente, G.** (1997) *Population Issues and the Situation of Women in Post-Conflict Guatemala*, ILO Action Programme on Skills and Entrepreneurship Training for Countries Emerging from Armed Conflict, Working Paper, International Labour Organisation, Geneva

LWF (1997) *Environmental Guidelines for Programme Implementation*, Department for World Service, Lutheran World Federation, Geneva, first version, March 1997

Mackintosh, A. (1996) Book review: The International Response to Conflict and Genocide: Lessons from the Rwanda Experience, Report of the Joint Evaluation of Emergency Assistance to Rwanda, *Journal of Refugee Studies* **9**(3): 334–42

Mamdani, M. (1996) From conflict to consent as the basis of state formation: reflections on Rwanda, *New Left Review*, **216**: 3–36

Mangel, M. et al. (1996) Principles for the conservation of wild living resources, *Ecological Applications*, **6**(2): 338–62

Manz, B. (1988) *Repatriation and Reintegration: An Arduous Process in Guatemala*. Hemispheric Migration Project, Center for Immigration Policy and Refugee Assistance, Georgetown University, Washington, DC

Markandya, A. and **Richardson, J.** (eds) (1992) *The Earthscan Reader in Environmental Economics*, Earthscan, London

Matose, F. (1997) Conflicts around forest reserves in Zimbabwe: what prospects for community management? *IDS Bulletin*, **28**(4): 69–78

Mattson, J. O. and **Rapp, A.** (1991) The recent droughts in western Ethiopia and Sudan in a climatic context, *Ambio*, **20**: 172–5

Mbodj, M., Sarr, M. and **Fall, S.** (1995) *Mission FAO d'Évaluation à Mi-parcours du Projet HCR OFADEC de Développement de Périmetres Irrigués pour l'Insertion Économique des Réfugiés Mauritaniens dans la Vallée du Fleuve Sénégal (du 16 au 30 janvier 1995)*, United Nations High Commissioner for Refugees, Geneva, and Food and Agriculture Organisation of the United Nations, Rome

McCallum, H. (1991) From destruction to recovery, in M. Ramphale and C. McDowell (eds) *Restoring the Land: Environment and Change in Post-Apartheid South Africa*, Panos Institute, London, pp. 167–76

McGregor, J. (1993) Refugees and the environment, in R. Black and V. Robinson (eds) *Geography and Refugees: Patterns and Processes of Change*, Belhaven, London, pp. 157–70

McGregor, J. (1997) *Staking their Claims: Land Disputes in Southern Mozambique*, LTC Paper No. 158, Land Tenure Centre, University of Wisconsin, Madison

McGregor, J., Harrell-Bond, B. and **Mazur, R.** (1991) *Mozambicans in Swaziland: Livelihood and Integration*, report by Refugee Studies Programme, University of Oxford, for World Food Programme, Rome

McSpadden, L. A. (1998) Contradictions and control in repatriation: negotiations for the return of 500 000 Eritrean refugees, in R. Black and K. Koser (eds) *The End of the Refugee Cycle: Refugee Repatriation and Reconstruction*, Berghahn, Oxford, in press

Meadows, D. H., Meadows, D. L., Behrens III, W. W. et al. (1972) *Limits to Growth*, Report of the Club of Rome, Universe Books, New York

Mearns, R. (1996) Environmental entitlements: pastoral natural resource management in Mongolia, *Cahiers des Sciences Humaines*, **32**(1): 105–31

Mercer, D. E. and **Soussan, J.** (1992) Fuelwood problems and solutions, in N. P. Sharma (ed.) *Managing the World's Forests: Looking for Balance between Conservation and Development*, World Bank, Washington, DC

Michel, J. H. (1996) *Development Co-operation: Efforts and Policies of the Members of the Development Assistance Committee*, Organisation for Economic Co-operation and Development, Paris

Millington, A. (1992) Soil erosion and conservation, in A. M. Mannion and S. R. Bowlby (eds) *Environmental Issues in the 1990s*, Wiley, Chichester, pp. 227–44

M'manga, W. (1991) Implications of population growth on land availability in Malawi, *Malawian Geographer*, **29**: 78–98

Morgan, W. B. and **Moss, R. P.** (1981) *Woodfuel and Rural Energy Production and Supply in the Humid Tropics, A Report for the United Nations University with Special Reference to Tropical Africa and South-East Asia*, Tycooly, Dublin, for the United Nations University

Mortimore, M. (1989) *Adapting to Drought: Farmers, Famines and Desertification in West Africa*, Cambridge University Press, Cambridge

Murphree, M. (1991) *Communities as Institutions for Resource Management*, CASS Occasional Paper Series, Centre for Applied Social Sciences, University of Zimbabwe, Harare

Murphree, M. and **Metcalfe, S. C.** (1997) *Conservancy Policy and the Campfire Programme in Zimbabwe*, CASS Occasional Paper Series, Centre for Applied Social Sciences, University of Zimbabwe, Harare

Mwale, P. E. S. and **Shaba, M. W. M.** (1996) *Evaluation Report on Multi-Disciplinary Environmental Rehabilitation Programme. The Dedza Refugee Impact Case*, Coordination Unit for the Rehabilitation of the Environment, Blantyre, July 1996

Myers, G. W. (1994) Competitive rights, competitive claims: land access in post-war Mozambique, *Journal of Southern African Studies*, **20**(4): 603–32

Myers, N. (1991) Tropical forests: present status and future outbook, *Climate Change*, **19**: 3–32

Myers, N. (1993a) Environmental refugees in a globally warmed world, *Bioscience*, **43**: 752–61

Myers, N. (1993b) *Ultimate Security: The Environmental Basis of Political Stability*, W.W. Norton, New York and London

Myers, N. (1993c) Tropical forests: the main deforestation fronts, *Environmental Conservation*, **20**(1): 9–16

Myers, N. (1993d) How many migrants for Europe? *People and the Planet*, **2**(3): 28

Myers, N. (1993e) Population, environment and development, *Environmental Conservation*, **20**(3): 205–14

Myers, N. (1996) Environmentally-induced displacements: the state of the art, in *Environmentally-Induced Population Displacements and Environmental Impacts Resulting from Mass Migration*, International Symposium, Geneva, 21–24 April 1996, International Organisation for Migration with United Nations High Commissioner for Refugees and Refugee Policy Group, pp. 72–3

Myers, N. and **Kent, J.** (1995) *Environmental Exodus: An Emergent Crisis in the Global Arena*, The Climate Institute, Washington, DC

Nauphal, N. (1997) *Post-War Lebanon: Women and Other War-Affected Groups*, ILO Action Programme on Skills and Entrepreneurship Training for Countries Emerging from Armed Conflict, Working Paper, International Labour Organisation, Geneva

Neefjes, K. and **David, R.** (1996) A *Participatory Review of the Ikafe Refugee Programme*, report for Oxfam UK&I, Oxford, October 1996

Nelson, H. D., Dobert, M., McLaughlin, J., Marvin, B. and **Whitaker, D. P.** (1975) *Area Handbook for Guinea*, 2nd edition, Foreign Area Studies Division, American University, Washington, DC

Neumann, R. P. and **Schroeder, R. A.** (1995) Manifest ecological destinies: local rights and global environmental agendas, *Antipode*, **27**(4): 321–4

Nhira, C. and **Matose, F.** (1995) *Lessons for the Resource Sharing Project in Zimbabwe from the Indian Joint Forest Management and Campfire Programme Experiences*, CASS Occasional Paper Series, Centre for Applied Social Sciences, University of Zimbabwe, Harare

Nimpuno, K. (1996) Emergency settlement, unsustainable development, in *Environmentally-Induced Population Displacements and Environmental Impacts Resulting from Mass Migration*, International Symposium, Geneva, 21–24 April 1996, International Organisation for Migration with United Nations High Commissioner for Refugees and Refugee Policy Group, pp. 89–91

NRI (1994) *Forestry Activities in Support of the Rehabilitation of Areas Affected by Refugees*, Mission Report, Natural Resources Institute, London.

NRI/ODA (1995) *Tanzania: The Environmental Situation in Kagera Region. Options for Future IFAD Interventions*, Natural Resources Institute/Overseas Development Administration, London

Nyoni, S. (1992) Women in stove programmes, *Boiling Point*, **27**: 9–11

ODA (1995) *A Guide to Social Analysis for Projects in Developing Countries*, Overseas Development Administration, London

OECD (1997) *DAC Guidelines on Conflict, Peace and Development Co-operation*, Organisation for Economic Co-operation and Development, Development Assistance Committee, Paris

Ofori-Cudjoe, S. (1990) Environmental Impact Assessment in Ghana: an ex post evaluation of the Volta Resettlement Scheme: the case of the Kpong Hydro-Electric Project, *The Environmentalist*, **10**(2): 115–26

O'Keefe, P., Kirkby, J. and **Cherrett, I.** (1991) Mozambican environmental problems: myths and realities, *Public Administration and Development*, **11**: 307–24

Orr, B. (1985) *Refugee Forestry in Somalia: The Step Plan Generates Community Involvement*, ICR Staff Paper Series No. 22, Inter Church Response for the Horn of Africa, Madison, Wisconsin

Otunnu, O. (1992) Environmental refugees in sub-Saharan Africa: causes and effects. *Refuge*, **12**(1): 11–14

Owen, M. (1996a) Developing sustainable household energy supply strategies in refugee-hosting areas, in *Refugees and the Environment in Africa*, Proceedings of a workshop at Bahari Beach, Dar-es-Salaam, Tanzania,

2–5 July 1996, United Nations High Commissioner for Refugees, Geneva, pp. 79–91

Owen, M. (1996b) *Refugees and Energy in Kagera*, report prepared for the United Nations High Commissioner for Refugees and the European Union

Owen, M. and Muchiri, L. (1994) *Energy Conservation in Refugee Camps in Burundi*, report for the International Federation of Red Cross and Red Crescent Societies, Nairobi, November 1994

Owen, M., Tiwari, D. and Coerver, A. (1998a) *Towards Sustainable Environmental Management in Refugee-Affected Areas: Country Report for Nepal*, draft report, Environment Unit, United Nations High Commissioner for Refugees, Geneva

Owen, M., Nyale, G. and Openshaw, K. (1998b) *Towards Sustainable Environmental Management in Refugee-Affected Areas: Country Report for Kenya*, draft report, Environment Unit, United Nations High Commissioner for Refugees, Geneva

Paquet, C. and van Soest, M. (1994) Mortality and malnutrition among Rwandan refugees in Zaire, *The Lancet*, 344: 823–4

Parker, T., Jimeleh, A., Ahmed, M., Okafo, O. and Abokar, F. (1987) *Hiran Refugee Reforestation Project: Final Evaluation, 16 February 1987*, CDA Forestry Project Grant No. 649–0122, CARE/USAID, Washington, DC

Parry, J. T. (1996) Radar remote sensing: land degradation and hazard monitoring, in M. J. Eden and J. T. Parry (eds) *Land Degradation in the Tropics: Environment and Policy Issues*, Pinter, London, pp. 19–38

Pearce, D., Hamilton, K. and Atkinson, G. (1996) Measuring sustainable development: progress on indicators, *Environment and Development Economics*, 1: 85–101

Peluso, N. L. (1995) Whose woods are these? Counter-mapping forest territories in Kalimantan, Indonesia, *Antipode*, 27(4): 383–406

Picard, M. (1996) *A Guide to the Gender Dimension of Environment and Natural Resources Management Based on a Sample Review of USAID NRM Projects in Africa*, Office of Sustainable Development, USAID, Washington, DC

Pimbert, M. (1997) Issues emerging in implementing the Convention on Biological Diversity, *Journal of International Development*, 9(3): 415–25

Pitterman, S. (1984) A comparative survey of two decades of international assistance to refugees in Africa, *Africa Today*, 31: 25–54

Place, F. and Otsuka, K. (1997) *Population, Land Tenure and Natural Resource Management: The Case of Customary Land in Malawi*, International Food Policy Research Institute, Washington, DC

Pottier, J. (1993) Migration as a hunger-coping strategy: paying attention to gender and historical change, in H. S. Marcussen (ed.) *Institutional*

Issues in Natural Resources Management, International Development Studies Occasional Paper No. 9, Roskilde University, Denmark, pp. 201–33

Pottier, J. (1996) Relief and repatriation: views by Rwandan refugees; lessons for humanitarian aid workers, *African Affairs,* **95**: 403–29

Pottier, J. (1998) The 'self' in self-repatriation: closing down Mugunga camp, Eastern Zaire, in R. Black and K. Koser (eds) *The End of the Refugee Cycle: Refugee Repatriation and Reconstruction,* Berghahn, Oxford, in press.

Pretty, J. N. and **Guijt, I.** (1992) Primary Environmental Care: an alternative paradigm for development assistance, *Environment and Urbanisation,* **4**(1): 22–36

Prunier, G. (1995) *The Rwandan Crisis 1959–1994: The History of a Genocide,* Hurst, London

Ramlogan, R. (1996) Environmental refugees: an overview, *Environmental Conservation,* **23**(1): 81–8

Redclift, M. (1984) *Development and the Environmental Crisis: Red or Green Alternatives?* Methuen, London

Reed, W. C. (1996) Exile, reform and the rise of the Rwandan Patriotic Front, *Journal of Modern African Studies,* **34**(3): 439–501

Richards, P. (1996) *Fighting for the Rain Forest: War, Youth and Resources in Sierra Leone,* International African Institute and James Currey, London

Rihoy, E. (1995) *The Commons Without the Tragedy? Strategies for Community-Based Natural Resource Management in Southern Africa. Proceedings of the Regional Natural Resource Management Programme Annual Conference, Kasane, Botswana, 3–6 April 1995,* USAID Regional Natural Resource Management Programme, Harare

Roberts, A. (1992) Destruction of the environment during the 1991 Gulf War, *International Review of the Red Cross,* **291**: 538–53

Roddick, J. (1997) Earth summit north and south: building a safe house in the winds of change, *Global Environmental Change,* **7**(2): 147–65

Roder, W. (1973) Effects of guerilla war in Angola and Mozambique, *Antipode,* **5**(2): 14–21

Ross, B. (1995) Cooking energy as seen by a planner, *Boiling Point,* **37**: 10–12

Ross, B. (1997) *Environmentally-Friendlier Procurement Guidelines,* Environment Unit, United Nations High Commissioner for Refugees, Geneva

Rouchiche, S. (1996) Environmental rehabilitation in refugee areas: case of Afghan refugees in Eastern Iran, in IOM (ed.) *Environmentally-Induced Population Displacements and Environmental Impacts Resulting from Mass Migration,* International Symposium, Geneva, 21–24 April 1996, International Organisation for Migration with United Nations High Commissioner for Refugees and Refugee Policy Group, pp. 104–6

Rounce, N. V. (ed.) (1949) *The Agriculture of the Cultivation Steppe of the Lake*, Western and Central Province, Longmans Tanganyika Territory, Dar es Salaam

RPN (1991) Avoiding camps, special issue of *Refugee Participation Network*, No. 10, Refugee Studies Programme, Oxford

Runge, C. (1984) Institutions and the free rider: the assurance problem in collective action, *Journal of Politics*, **46**: 154–81

Russell, S., Jacobsen, K. and **Stanley, W. D.** (1990) *International Migration and Development in Sub-Saharan Africa*, two volumes, World Bank, Washington, DC

Rutinwa, B. (1996) The Tanzanian government's response to the Rwandan emergency, *Journal of Refugee Studies*, **9**(3): 291–302

Sabiti, M. (1996) Afforestation project in former refugee-hosting areas in Malawi, in *Refugees and the Environment in Africa*, Proceedings of a workshop at Bahari Beach, Dar-es-Salaam, Tanzania, 2–5 July 1996, United Nations High Commissioner for Refugees, Geneva, pp. 32–3

Sahabat Alam Malaysia (1992) Environmental displacement in Malaysia: the effects of the development process on rural and native communities, *Refuge*, **12**(1): 15–19

Salem-Murdock, M., Horowitz, M. and **Scudder, T.** (1989) *La Suivi des Activités Agricoles dans la Moyenne Vallée du Sénégal: Progrès des Recherches*, Institute for Development Anthropology, Binghampton, NY

Sanders, T. G. (1990–91) *Northeast Brazilian Environmental Refugees: Where They Go, Parts I and II*, Field Staff Report No. 21, Universities Field Staff International, Washington, DC

Sane, A. (1993) *La Population Réfugiée au Sénégal*. Memoire de Maîtrisse, Département de Géographie, Université Cheikh Anta Diop, Dakar

Sawyer, J. (1990) Tropical forests: multiple use, not multiple abuse, *Earthwatch*, **39**: 4–6

Sayer, A. (1982) Explanation in economic geography: abstraction versus generalization, *Progress in Human Geography*, **6**(1): 68–88

Schwartz, M. L. and **Notini, J.** (1995) Preliminary report on desertification and migration: case studies and evaluation, in J. Puigdefábrigas and J. Mendizábal (eds) *Desertification and Migrations*, Geoforma Ediciones, Logroño, Spain, pp. 69–113

Scoones, I. (ed.) (1995) *Living with Uncertainty: New Directions in Pastoral Development in Africa*, Intermediate Technology, London

Scoones, I. (1996) Range management science and policy: politics, polemics and pasture in southern Africa, in M. Leach and R. Mearns (eds) *The Lie of the Land: Challenging Received Wisdom on the African Environment*, International African Institute and James Currey, London, pp. 34–53

Seldman, A. and **Seldman, R. B.** (1995) *Participatory Resource Analysis and the Creation of a Democratic Law Making Process,* African Studies Center Working Paper No. 197, Boston University, Boston, Mass.

Sen, A. (1981) *Poverty and Famines,* Clarendon, Oxford

Sepp, S., Seibel, S. and **Munyarugerero, G.** (1995) *Natural Resources Rehabilitation in the Area Surrounding the Rwandan Refugee Camps, Kagera Region,* project appraisal report, Deutsche Gesellschaft für Technische Zusammenarbeit GmbH, Oberaula, Germany

Shackley, S. (1997) The Intergovernmental Panel on Climate Change: consensual knowledge and global politics, *Global Environmental Change,* 7(1): 77–9

Shapiro, D. (1989) *Population Growth, Urbanisation and Changing Agricultural Practices in Zaire,* Working Paper, Population Research Institute, Pennsylvania State University, University Park, Pa.

Shaw, R. P. (1993) Warfare, national sovereignty and the environment, *Environmental Conservation,* 20(2): 113–21

Sheckler, A. C., Tardif-Douglin, D., Flouriot, J., Adisa, J., Loosi, E., Lanjouw, A., Berry, L. and **Ekoko, F.** (1996) *Strategic Action Plan for the Great Lakes Region of Africa (Rwanda, Burundi, Tanzania, Uganda, Zaire),* United Nations Environment Programme and United Nations Centre for Human Settlements (Habitat), Nairobi

Siddique, A. K., Salam, A., Islam, M. S., Akram, K., Majumdar, R. N., Zaman, K., Fronczak, N. and **Laston, S.** (1995) Why treatment centres failed to prevent cholera deaths among Rwandan refugees in Goma, Zaire, *The Lancet,* 345: 359–61

SNSA (1996) *Étude d'Impact des Aménagements de Bas Fonds PNIR et Programmes Associés,* two volumes, Service Nationale des Statistiques Agricoles, Ministère de l'Agriculture, Eaux et Forêts, Republic of Guinea, Conakry

Sorenson, J. (1994) An overview: refugees and development, in H. Adelman and J. Sorenson (eds) *African Refugees: Development Aid and Repatriation,* Westview, Boulder, Colo., pp. 175–90

Sparrow, A. (1993) *Indigenous Woodland Resources: Chambuta Refugee Camp,* unpublished report to the Fuelwood Crisis Consortium, Harare, Zimbabwe

Spellerberg, I. F. (1992) *Evaluation and Assessment for Conservation,* Chapman & Hall, London

Spitteler, M. (1992) *Tree Planting: Refugee and Host Community Needs. A Case Study of Ukwimi Refugee Settlement, Zambia,* unpublished report, Refugee Studies Programme, Oxford

Spooner, B. (1987) Insiders and outsiders in Baluchistan: western and indigenous perspectives on ecology and development, in P. Little and

M. Horowitz (eds) *Lands at Risk in the Third World: Local Level Perspectives*, Westview, Boulder, Colo., pp. 58–68

Squires, V. (1994) *IFAD/UNHCR/IRI South Khorosan Rangeland Rehabilitation and Afghan Refugee Income Generating Project, Aide-Memoire*, International Fund for Agricultural Development, Rome

Ståhl, M. (1990) *Constraints to Environmental Rehabilitation through People's Participation in the Northern Ethiopian Highlands*, UNRISD Discussion Paper No. 13, United Nations Research Institute for Social Development, Geneva

Stark, O. and **Lucas, R. E. B.** (1988) Migration, remittances and the family, *Economic Development and Cultural Change*, **36**: 465–81

Steinbeck, J. (1939) *The Grapes of Wrath*. Heinemann, London

Stockton, N. (1996) The role of the military in humanitarian emergencies: reflections, *Refugee Participation Network*, **23**: 13–18

Suhrke, A. (1993) *Pressure Points: Environmental Degradation, Migration and Conflict*, occasional paper of Project on Environmental Change and Acute Conflict, American Academy of Arts and Sciences, Washington, DC

Sullivan, S. (1990) *Common Property Resources Utilized at Ukwimi Mozambican Agricultural Settlement, Eastern Province, Zambia*, unpublished report, lodged with the Royal Geographical Society, London

Swift, A. (1993) *Global Political Ecology: The Crisis in Economy and Government*, Pluto, London

Swift, J. (1996) Desertification: narratives, winners and losers, in M. Leach and R. Mearns (eds) *The Lie of the Land: Challenging Received Wisdom on the African Environment*, International African Institute and James Currey, London, pp. 73–90

Tabor, J. A. and **Hutchinson, C. F.** (1994) Using indigenous knowledge, remote sensing and GIS for sustainable development, *Indigenous Knowledge and Development Monitor*, **2**(1): 2–6

Tamondong-Helin, S., and **Helin, W.** (1991) *Migration and the Environment: Interrelationships in Sub-Saharan Africa*: Field Staff Reports, No. 22, Universities Field Staff International, Washington, DC

Thanh, N. C. and **Tam, D. M.** (1992) Environmental protection and development: how to achieve a balance? in A. K. Biswas and S. B. C. Agarwala (eds) *Environmental Impact Assessment for Developing Countries*, Butterworth-Heinemann, Oxford, pp. 3–15

Thérivel, R. and **Partidário, M. R.** (1996) *The Practice of Strategic Environmental Assessment*, Earthscan, London

Thiadens, R. (1992) *Senegal: Self-Sufficiency Programme for Mauritanian Refugees*, PTSS Mission Report 92/54, United Nations High Commissioner for Refugees, Geneva

Thiadens, R. and **Mori, H.** (1995) Environmental issues: UNHCR's experience and response, in *The Environmental Impact of Sudden Population Displacements: Priority Policy Issues and Humanitarian Aid*, expert consultation sponsored by ECHO and organised by CRED, Université Catholique de Louvain, Brussels, pp. 28–35

Thomas, D. S. G. and **Middleton, N. J.** (1994) *Desertification: Exploding the Myth*, Wiley, Chichester

Thornes, J. B. and **Brunsden, D.** (1977) *Geomorphology and Time*, Methuen, London

Tiffen, M., Mortimore, M. and **Gichuki, F.** (1994) *More People, Less Erosion: Environmental Recovery in Kenya*, Wiley, Chichester

Timberlake, L. (1988) *African Crisis: The Causes, the Cures of Environmental Bankruptcy*, Earthscan, London

Tol, R. S. J. (1995) The damage costs of climate change: towards more comprehensive calculations, *Environmental and Resource Economics*, **5**: 353–74

Toussaint, A., Ducenne, Q. and **Roulette, G.** (1994) *Projet de Restauration du Milieu Naturel dans le Département de Podor, République de Sénégal. Rapport de la Première Phase*, Deutsche Forstservice GmbH, Feldkirchen, Germany, for Direction des Eaux, Forêts et de la Conservation des Sols, République de Sénégal, Dakar

Trolldalen, J. M., Birkeland, N., Borgen, J. and **Scott, P. T.** (1992) *Environmental Refugees: A Discussion Paper*, World Foundation for Environment and Development and Norwegian Refugee Council, Oslo

Tucker, C. J., Dregne, H. E. and **Newcomb, W. W.** (1991) Expansion and contraction of the Sahara Desert from 1980 to 1990, *Science*, **253**: 299–301

Tuite, P. and **Gardiner, J. J.** (1990) The miombo woodlands of Central, Eastern and Southern Africa, *Irish Forestry*, **47**(2): 90–107

Turton, D. (1988) Looking for a cool place: the Mursi, 1890s–1980s, in D. H. Johnson and D. M. Anderson (eds) *The Ecology of Survival: Case Studies from Northeast African History*, Westview, Boulder, Colo.

Turton, D. and **Turton, P.** (1984) Spontaneous resettlement after drought: a Mursi case study, *Disasters*, **8**(3): 178–89

Umlas, E. (1996a) *The Experience of UNHCR and its Partners with Solar Cookers in Refugee Camps*, Office of the Senior Coordinator on Environmental Affairs, United Nations High Commissioner for Refugees, Geneva

Umlas, E. (1996b) Household energy use in refugee camps of Eastern Zaire and Tanzania. The experiences of UNHCR and its implementing partners, *Boiling Point*, **37**: supplement

UNECA (1991) *Measures for Villagisation and Resettlement in Selected African Countries*, United Nations Economic Commission for Africa, Addis Ababa

UNEP (1992) *World Atlas of Desertification*, Edward Arnold, London

UNHCR (1982) *Handbook for Emergencies*, United Nations High Commissioner for Refugees, Geneva

UNHCR (1991) Some environmental considerations in refugee camps and settlements, *Rapport*, **10**: 1–4

UNHCR (1996) *Environmental Guidelines*, Office of the Senior Coordinator on Environmental Affairs, United Nations High Commissioner for Refugees, Geneva

UNHCR (1997) *Refugees and Others of Concern to UNHCR, 1996 Statistical Overview*, Office of the United Nations High Commissioner for Refugees, Geneva

UNHCR/GTZ (1995) *Bulletin d'Information sur l'Environnement*, No. 3, Goma, October 1995

URT (1995) *Assessment Report on the Impact of Refugees on the Local Communities in Kagera and Kigoma Regions. Part 1: Kagera Region*, unpublished report, The United Republic of Tanzania Office of the Prime Minister and First Vice President

USAID (1990) *Senegal Agricultural Sector Analysis*, United States Agency for International Development, Dakar

USCR (1995) *The Return of Rohingya Refugees to Burma: Voluntary Repatriation or Refoulement?* Issue Paper, United States Committee for Refugees, Washington, DC

USCR (1997) *World Refugee Survey 1997*, United States Committee for Refugees, Washington, DC

Van Damme, W. (1995) Do refugees belong in camps? Experiences from Goma and Guinea, *The Lancet*, **346**: 360–2

Van Damme, W. (1997) *The Refugee Assistance Programme in Guinea (1990–1996): An Alternative Approach Avoiding Encampment. From Collective Wishful-Thinking to Peaceful Co-habitation?* unpublished paper, Department of Public Health, Institute for Tropical Medicine, Antwerp

Van den Breemer, H. P. M. and Venema, L. B. (1995) Local resource management in African national contexts, in J. P. M. van den Breemer, C. A. Drijver and L. B. Venema (eds) *Local Resource Management in Africa*, Wiley, Chichester, pp. 3–25

Van den Breemer, J. P. M., Bergh, R. R. and Hesseling, G. (1995) Towards local management of natural resources in Senegal, in J. P. M. van den Breemer, C. A. Drijver and L. B. Venema (eds) *Local Resource Management in Africa*, Wiley, Chichester, pp. 97–110

Van der Borght, S. and Philips, M. (1995) Do refugees belong in camps? *The Lancet*, **346**: 907–8

Van Lavieren, B. and van Wetten, J. (1990) *Profil de l'Environnement de la Vallée du Fleuve Sénégal*, Euroconsult/Institut National de Recherche pour la Conservation de la Nature, Dakar

217

Van Pelt, M. J. F. (1993) *Ecological Sustainability and Project Appraisal. Case Studies in Developing Countries*, Avebury, Aldershot

Velenchik, A. D. (1993) Cash-seeking behaviour and migration: a place-to-place migration function for Côte d'Ivoire, *Journal of African Economies*, 2(3): 329–47.

Vincent, B. (1995) *Assistance aux Réfugiés, Goma Zaïre: Secteur Économie de Bois de Feu. Rapport de Mission (II) du 22.11.94 au 05.01.95*, Deutsche Gesellschaft für Technische Zusammenarbeit, Goma

Wagner, R. H. (1974) Indochina: the war against an environment, in R. H. Wagner (ed.) *Environment and Man*, W. W. Norton, New York, pp. 360–74

Waldron, S. and **Hasci, N. A.** (1995) *Somali Refugees in the Horn of Africa. State of the Art Literature Review*, Studies on Emergencies and Disaster Relief, Report No. 3, Nordiska Afrikainstitutet, Uppsala, Sweden

Walker, D. (1989) *Famine Early Warning Systems: Victims and Destitution*, Earthscan, London

Waller, D. (1993) *Rwanda: Which Way Now?* Oxfam, Oxford

Walsh, M. (1997) *Post-Conflict Bosnia and Herzegovina: Integrating Women's Special Situation and Gender Perspectives in Skills Training and Employment Promotion Programmes*, ILO Action Programme on Skills and Entrepreneurship Training for Countries Emerging from Armed Conflict, Working Paper, International Labour Organisation, Geneva

Wandira, K. and **Hoerz, T.** (1997) *Participatory Environmental Appraisal and Planning (PEAP) for Maji and Mongula Settlement, Sub-Office Pakelle*, unpublished report, Mitigation of Environmental Degradation in Refugee Hosting Areas through Participatory Systems Research Project, GTZ, Eschborn, Germany

Wathern, P. (1988) An introductory guide to EIA, in P. Wathern (ed.) *Environmental Impact Assessment: Theory and Practice*, Routledge, London and New York, pp. 1–30

Watts, M. (1991) Entitlements or empowerment? Famine and starvation in Africa, *Review of African Political Economy*, 51: 9–26

WCED (1987) *Our Common Future*, report of the World Commission on Environment and Development, Oxford University Press, Oxford

Welford, R. (1997) *Hijacking Environmentalism: Corporate Responses to Sustainable Development*, Earthscan, London

Werner, G. (1992) EIA in Asia, in A. K. Biswas and S. B. C. Agarwala (eds) *Environmental Impact Assessment for Developing Countries*, Butterworth-Heinemann, Oxford, pp. 16–22

Westing, A. (1976) *Ecological Consequences of the Second Indochina War*, Almqvist and Wiksell, Stockholm

Westing, A. (1992) Environmental refugees: a growing category of displaced persons, *Environmental Conservation*, **19**(3): 201–7

WFP (1992) *Food Aid Review*, World Food Programme, Rome

Whitney, J. (1992) The Three Gorges Project in China, *Refuge*, **12**(1): 24–6

Wijkman, A. and **Timberlake, L.** (1984) *Natural Disasters*, Earthscan, London

Wilkie, D. S. (1994) Remote sensing imagery for resource inventories in Central Africa: the importance of detailed field data, *Human Ecology*, **22**(3): 379–404

Williams, M. A. J. and **Balling Jr, R. C.** (1996) *Interactions of Desertification and Climate*, Edward Arnold, London

Wilson, E. O. and **Peters, F. M.** (eds) (1988) *Biodiversity*, National Academy Press, Washington, DC

Wilson, K. (1992) *A State of the Art Review of Research on Internally Displaced, Refugees and Returnees from and in Mozambique*, report prepared for the Swedish International Development Authority, June 1992

Wilson, K. (1995) Alleviating deforestation around refugee camps in the African context, in C. Nhira (ed.) *Proceedings of the Conference on 'Indigenous Woodland Management'*, Centre for Applied Social Sciences, University of Zimbabwe, Harare, 22–24 February 1994

Wilson, K., Cammack, D. and **Shumba, F.** (1989) *Food Provisioning amongst Mozambican Refugees in Malawi: A Study of Aid, Livelihood and Development*, report for World Food Programme by the Refugee Studies Programme, Oxford

Witte, J. (1992) Deforestation in Zaire: logging and landlessness, *The Ecologist*, **22**(2): 58–64

Woodhouse, P. and **Ndiaye, I.** (1990) *Structural Adjustment and Irrigated Food Farming in Africa: the 'Disengagement' of the State in the Senegal River Valley*, DPP Working Paper No. 20, The Open University, Milton Keynes

World Bank (1996a) *Pakistan Impact Evaluation Report, June 28, 1996*, Report No. 15862-PAK, Operations Evaluation Department, World Bank, Washington, DC

World Bank (1996b) *Malawi's Human Resources and Poverty: Profile and Priorities for Action*, World Bank, Washington, DC

Yeraswork, A. (1988) *Impact and Sustainability of Activities of Rehabilitation of Forest, Grazing and Agricultural Lands supported by World Food Programme Project 2488*, Report to WFP and Natural Resources Main Department, Ministry of Agriculture, Addis Ababa

Young, L. (1985) A general assessment of the environmental impact of refugees in Somalia, with attention to the refugee agricultural programme, *Disasters*, **9**: 122–33

Zabala, H. (1994) *Physical planning activities in the Uvira and Bukavu regions, South Kivu, Eastern Zaire, 4 May–4 November 1994*, PTSS Mission Report 94/61, United Nations High Commissioner for Refugees, Geneva

Zaman, M. Q. (1989) The social and political context of adjacent to riverbank erosion hazard and population resettlement in Bangladesh, *Human Organisation*, **48**(3): 196–205

Zetter, R. (1988) Refugees and refugee studies: a label and an agenda, *Journal of Refugee Studies*, **1**(1): 1–6

Zetter, R. (1995) *Shelter Provision and Settlement Policies for Refugees: A State of the Art Review*, Studies on Emergencies and Disaster Relief No. 2, Nordiska Afrikainstitutet, Uppsala

Index

Note: References in endnotes are denoted by *n* after the page number.